Big Data Management in Sensing: Applications in AI and IoT

RIVER PUBLISHERS SERIES IN BIOMEDICAL ENGINEERING

Series Editors:

DINESH KANT KUMAR
RMIT University, Australia

The "River Publishers Series in Biomedical Engineering" is a series of comprehensive academic and professional books which focus on the engineering and mathematics in medicine and biology. The series presents innovative experimental science and technological development in the biomedical field as well as clinical application of new developments.

Books published in the series include research monographs, edited volumes, handbooks and textbooks. The books provide professionals, researchers, educators, and advanced students in the field with an invaluable insight into the latest research and developments.

Topics covered in the series include, but are by no means restricted to the following:

- Biomedical engineering
- Biomedical physics and applied biophysics
- Bio-informatics
- Bio-metrics
- Bio-signals
- Medical Imaging

For a list of other books in this series, visit www.riverpublishers.com

Big Data Management in Sensing: Applications in AI and IoT

Editors

Renny Fernandez
Norfolk State University, USA

Terrance Frederick Fernandez
Dhanalakshmi Srinivasan College of Engineering & Technology, India

LONDON AND NEW YORK

Published 2021 by River Publishers
River Publishers
Alsbjergvej 10, 9260 Gistrup, Denmark
www.riverpublishers.com

Distributed exclusively by Routledge
4 Park Square, Milton Park, Abingdon, Oxon OX14 4RN
605 Third Avenue, New York, NY 10017, USA

Big Data Management in Sensing: Applications in AI and IoT / by Renny Fernandez, Terrance Frederick Fernandez.

© 2021 River Publishers. All rights reserved. No part of this publication may be reproduced, stored in a retrieval systems, or transmitted in any form or by any means, mechanical, photocopying, recording or otherwise, without prior written permission of the publishers.

Routledge is an imprint of the Taylor & Francis Group, an informa business

ISBN 978-87-7022-415-4 (print)

While every effort is made to provide dependable information, the publisher, authors, and editors cannot be held responsible for any errors or omissions.

Contents

Preface	xv
List of Figures	xvii
List of Tables	xxi
List of Contributors	xxiii
List of Abbreviations	xxvii

1 Classification of Histopathological Variants of Oral Squamous Cell Carcinoma Using Convolutional Neural Networks   1
P. Archana, T. Megala, D. Udaya, and S. Prabavathy
- 1.1 Introduction 2
- 1.2 Convolutional Neural Networks 4
  - 1.2.1 Convolutional Layer 5
  - 1.2.2 Pooling Layer 5
  - 1.2.3 Fully Connected Layers 5
  - 1.2.4 Receptive Field 5
  - 1.2.5 Weights 6
  - 1.2.6 ReLU Layer 6
  - 1.2.7 Softmax Layer 6
  - 1.2.8 Dropout 6
  - 1.2.9 Steps Involved in Convolutional Neural Network .. 7
- 1.3 Proposed Convolutional Neural Network 7
  - 1.3.1 Performance Evaluation for CNN Models 8
  - 1.3.2 Comparative Result Analysis 10
- 1.4 Conclusion 12
- References 12

2 Voice Recognition Using Natural Language Processing — 15
J. Pradeep, K. Vijayakumar, and M. Harikrishnan
- 2.1 Introduction — 15
- 2.2 Proposed System — 17
  - 2.2.1 Automatic Speech Recognition — 17
  - 2.2.2 Auto-detect Language — 18
  - 2.2.3 Syntactic Analysis — 18
  - 2.2.4 Semantic Analysis — 18
  - 2.2.5 Pragmatic Analysis — 19
- 2.3 Experimental Results — 19
- 2.4 Conclusion — 22
- References — 22

3 Detection of Tuberculosis Using Computer-Aided Diagnosis System — 25
Murali Krishna Puttagunta, S. Ravi, and A. Anbarasi
- 3.1 Introduction — 26
- 3.2 Pre-Processing — 28
- 3.3 Segmentation — 28
  - 3.3.1 Rule-Based Algorithm — 28
  - 3.3.2 Pixel Classification — 29
  - 3.3.3 Deformable Models — 29
  - 3.3.4 Hybrid Methods — 30
- 3.4 Feature Extraction — 30
  - 3.4.1 Histogram Features — 30
  - 3.4.2 Shape Descriptor Histogram — 31
  - 3.4.3 Curvature Descriptor — 31
  - 3.4.4 Local Binary Pattern (LBP) — 31
  - 3.4.5 Histogram of Gradients — 32
  - 3.4.6 Gabor Features — 32
- 3.5 Classification — 33
- 3.6 Discussion — 34
- 3.7 Conclusion — 37
- References — 38

4 Forecasting Time Series Data Using ARIMA and Facebook Prophet Models — 47
S. Sivaramakrishnan, Terrance Frederick Fernandez, R. G. Babukarthik, and S. Premalatha
- 4.1 Introduction — 48

	4.2	Arima Model	50
		4.2.1 Data Analysis Using ARIMA Model	51
	4.3	Data Analysis Using Facebook Prophet Model	55
	4.4	Conclusion	57
		References	57

5 A Novel Technique for User Decision Prediction and Assistance Using Machine Learning and NLP: A Model to Transform the E-commerce System   61

V. Vivek, T. R. Mahesh, C. Saravanan, and K. Vinay Kumar

5.1	Introduction	62
5.2	Related Work	64
5.3	Research Methodology	68
5.4	Experimental Results	72
5.5	Conclusion and Future Scope	74
	References	75

6 Machine Learning-Based Intelligent Video Analytics Design Using Depth Intra Coding   77

Kumbala Pradeep Reddy, Sarangam Kodati, Thotakura Veeranna, and G. Ravi

6.1	Introduction		78
	6.1.1	Object Detection	80
	6.1.2	Deep Learning	80
	6.1.3	Geometric Depth Modeling	80
		6.1.3.1 Plane fitting	80
	6.1.4	Depth Coding Based on Geometric Primitives	81
6.2	Video Analytics Design Using Depth Intra Coding		82
6.3	Results		83
6.4	Conclusion		85
	References		85

7 A Novel Approach for Automatic Brain Tumor Detection Using Machine Learning Algorithms   87

G. Sindhu Madhuri, T. R. Mahesh, and V. Vivek

7.1	Introduction		88
	7.1.1	Medical Imaging	89
7.2	Image Processing Approach-Detection of Brain Tumor From Mri Images		90

viii Contents

	7.3	Machine Learning Approach-Detection of Brain Tumor From MRI Images	94
	7.4	Nano-Robotic Approach-Detection of Brain Tumor From Mri Images	98
		References	99

8 A Swarm-Based Feature Extraction and Weight Optimization in Neural Network for Classification on Speaker Recognition 103
G. Raja, P. Salini, M. Pradeep, and Terrance Frederick Fernandez

	8.1	Introduction	104
		8.1.1 Swarm-based Feature Extraction Merits	104
		8.1.2 Objectives of Our Chapter	105
	8.2	State of Art	105
		8.2.1 Mel Frequency Cepstral Coefficients (MFCC)	106
		8.2.2 Swarm Intelligence (SI)	106
		8.2.3 Text-independent Speaker Identification	106
		8.2.4 Voice Activity Detection (VAD)	107
	8.3	Differential Evolution Technique (DE)	107
	8.4	Survey on Swarm Intelligence	107
	8.5	Our Framework and Metrics	108
	8.6	Results and Discussion	110
		References	112

9 Fault Tolerance-Based Attack Detection Using Ensemble Classifier Machine Learning with IOT Security 115
A. Arulmurugan, R. Kaviarasan, and Saiyed Faiayaz Waris

	9.1	Introduction	116
	9.2	Background	118
		9.2.1 IoT Security Attacks	118
		9.2.1.1 Perception Layer Attacks	118
		9.2.1.2 Network Layer Attacks	119
		9.2.1.3 Routing Attacks	119
	9.3	Deep Learning and IoT Security	120
	9.4	Deep Learning and Big Data Technologies for IoT Security	123
	9.5	Cloud Framework for Profound Learning, Enormous Information Advances, and IoT Security	124
		9.5.1 Related Works	124

9.6		Motivation of the Proposed Methodology	126
9.7		Research Methodology	126
	9.7.1	Dimensionality Reduction	128
	9.7.2	Independent Component Analysis	129
	9.7.3	Principal Component Analysis	130
	9.7.4	Cloud Architecture	131
	9.7.5	Encryption Decryption Using OTP	131
	9.7.6	OTP Algorithm	135
	9.7.7	Ensemble Classifier SVM, Random Forest Classification	136
	9.7.8	Random Forest	139
9.8		Performance Metrics	140
9.9		Dataset Description	141
9.10		Conclusion	145
		References	146

10 Design a Novel IoT-Based Agriculture Automation Using Machine Learning 149

G. Ravi, Kumbala Pradeep Reddy, M. Mohan Rao,
Sarangam Kodati, and J. Praveen Kumar

10.1	Introduction	150
10.2	Literature Survey	151
10.3	Novel IoT-Based Agriculture Automation Using Machine Learning	153
10.4	Conclusion	156
	References	156

11 Building a Smart Healthcare System Using Internet of Things and Machine Learning 159

Shruti Kute, Amit Kumar Tyagi,
Rohit Sahoo, and Shaveta Malik

11.1	Smart Healthcare—An Introduction	160
11.2	Background Study	161
11.3	Motivation of This Work	162
11.4	Internet of Things–Enabled Safe Smart Hospital Cabin Door Knocker	162
11.5	Smart Healthcare System Communication Protocol	164
11.6	IoT-Cloud Based Smart Healthcare Data Collection System	165

- 11.7 Use of Machine Learning in Different Fields of Medical Science 166
- 11.8 Illness Identification/Diagnosis 167
  - 11.8.1 Discovery of Drug & Manufacturing 167
  - 11.8.2 Diagnosis of Medical Imaging 168
  - 11.8.3 Clinical Trial 168
  - 11.8.4 Epidemic Outbreak Prediction 168
  - 11.8.5 Robotic Surgery 168
  - 11.8.6 Smart Health Record 169
- 11.9 Challenge'S Faced Towards 5G With Iot and Machine Learning Technique 169
  - 11.9.1 5G and IoT Empower More Assault Vectors 169
  - 11.9.2 Smarter Bots Can Likewise Misuse These Assault Vectors 170
- 11.10 Future Possibility of Smart Healthcare With Internet of Things 171
- 11.11 Conclusion and Future Scope 173
  - References 174

12 Research Issues and Future Research Directions Toward Smart Healthcare Using Internet of Things and Machine Learning 179

Shruti Kute, Amit Kumar Tyagi, and Meghna Manoj Nair

- 12.1 Introduction 180
- 12.2 Background Work 180
- 12.3 Healthcare and Internet of Things 185
- 12.4 Internet of Things-Based Healthcare Solutions 185
  - 12.4.1 Clinical Care 186
  - 12.4.2 Distant Checking 186
- 12.5 Machine Learning-Based Healthcare 186
  - 12.5.1 Future Model of Healthcare-based IoT and Machine Learning 187
- 12.6 Wearable System for Smart Healthcare 189
- 12.7 Communication Standards 190
- 12.8 Challenges in Healthcare Adoption with IoT and Machine Learning 191
- 12.9 Improving Adoption of Healthcare System with IoT and Machine Learning 192
  - 12.9.1 Proof-based Consideration 192
  - 12.9.2 Self-learning and Personal Growth 193

		12.9.3 Normalization	194

        12.9.3 Normalization 194
        12.9.4 Protection and Security 194
        12.9.5 Intelligent Announcing and Representation 195
  12.10 Proposed Solution Based on IOT and Machine Learning
        for Smart Healthcare Systems 195
  12.11 Conclusion 198
        References 199

13 A Novel Adaptive Authentication Scheme for Securing Medical Information Stored in Clouds   201

N. Moganarangan, N. Palanivel, and S. Balaji

  13.1 Introduction 202
  13.2 Adaptive Authentication Scheme 204
  13.3 Information Storage/Update 205
  13.4 Integrity Check 208
  13.5 Performance Analysis 209
        13.5.1 Process Delay 209
        13.5.2 Integrity Check Bytes 210
        13.5.3 Overhead 210
  13.6 Conclusion 212
        References 212

14 E-Tree MSI Query Learning Analytics on Secured Big Data Streams   215

B. Balamurugan and S. Jegadeeswari

  14.1 Introduction 216
  14.2 Literature Review 217
  14.3 Proposed Framework-Secured Framework for Balancing
        Load Factor Using Ensemble Tree Classification 218
        14.3.1 Fast Predictive Look-ahead Scheduling Approach . 220
        14.3.2 Parallel Ensemble Tree Classification (PETC) . . . 221
        14.3.3 Bilinear Quadrilateral Mapping 222
  14.4 Conclusion 222
        References 223

15 Lethal Vulnerability of Robotics in Industrial Sectors   227

R. G. Babukarthik, Terrance Frederick Fernandez,
Sivaramakrishnan, and Aiswariya Milan

  15.1 Introduction 228

		15.1.1 Robotics' Impact on Manufacturing Industries	228
	15.2	Robotics and Innovation	228
		15.2.1 Data Collection	229
		15.2.2 Walking Robots	229
		15.2.3 Various Robot Names and Dimensions	230
	15.3	Robot Service in Hotels	231
		15.3.1 Study 1A	233
		15.3.2 Study 1B	233
	15.4	Cyber Security Attacks on Robotic Platforms	234
	15.5	Conclusion	235
		References	236

16 Smart IoT Assistant for Government Schemes and Policies Using Natural Language Processing — 239

J. Pradeep, K. Manojkiran, V.P. Gopi, and B. Jayakumar

16.1	Introduction		240
16.2	Literature Survey		240
16.3	Proposed Smart System		243
	16.3.1 Data Extraction		244
	16.3.2 Data Processing		244
	16.3.3 Sending SMS		245
	16.3.4 Language Translation		245
	16.3.5 Text-To-Speech		245
		16.3.5.1 Input text	246
		16.3.5.2 Text analysis	246
		16.3.5.3 Phonetic analysis	246
		16.3.5.4 Speech database	246
		16.3.5.5 Concatenation & Waveform generation	247
		16.3.5.6 Synthesized speech	247
16.4	Methodology		247
	16.4.1 Input Text Data		247
	16.4.2 URL Data Extraction		248
	16.4.3 Image to Text Conversion		248
	16.4.4 Extract Text from PDF		248
	16.4.5 SMS Update		249
	16.4.6 GSM		249
	16.4.7 Language Selection		249
	16.4.8 Text-To-Speech		249
	16.4.9 GUI		250

16.5	Experimental Results	250
16.6	Conclusion	252
	References	252

Index   255

About the Editors   257

Preface

The book *Big Data Management in Sensing: Applications in AI and IoT* places its focus upon enhancing the accuracy of the current Internet architecture that is driven by AI enabled approaches and IoTs. It features a wide range of research manuscripts and extensive survey upon Machine Learning algorithms and IoT applied to Medical diagnostics, Language processing, voice classification, social media forecasting, video analytics and smart handling of Government policies.

The principal motivation behind collating manuscripts for our book is due to a lack of information on Network optimization approaches powered by deep learning and IoT, although a plethora of journals publish on AI and IoTs. Secondarily, we drew inspiration from applications of deep learning in medical arenas for analysis, detection of tumors and the behavior of an epidemic spread like COVID, that terrorize globally now. Furthermore, there is a dire need to associate deep learning as a smart tool for classification and regression in video analytics and speech recognition.

The book is crafted for researchers, students and academicians in the field of data science, who wish to update their skills and comprehend the latest techniques in data analysis and visualization. Professionals across AI based industries in government agencies and businessmen in private sectors may be fascinated by the applications of blended deep learning models across medical arenas, digital recognition, and robots. Some of the chapters discuss implementation results carried out with mathematical models employing Orange data science toolkit, Weka tool and many more.

Mighty challenged by the worldwide outbreak of the pandemic and novelty of the book theme, we have compiled 16 chapters, nested under two sub-themes, from a divergent array of applications in AI and IoT. We expect the book to be a resourceful reference for academic researchers and professionals in the area of applied machine learning.

Renny Edwin Fernandez
Terrance Frederick Fernandez

List of Figures

Figure 1.1	Proposed framework of histopathological variants of OSCC.	3
Figure 1.2	Architecture of Convolution Neural Network.	4
Figure 1.3	Proposed frameworks of CNN.	7
Figure 1.4	Classification results of different CNN models.	11
Figure 2.1	Proposed method.	17
Figure 2.2	Framework of language technology and speech recognition.	18
Figure 2.3	NLP algorithm.	19
Figure 2.4	Input to the system.	20
Figure 2.5	Text document.	21
Figure 2.6	Output screen.	21
Figure 3.1	Chest radiography (CXR) with computer-aided detection (CAD) used as a triage test.	27
Figure 3.2	(a) Chest anatomy, (b) A healthy chest X-ray image, (c) Tuberculosis with multiple cavitation's CXR images.	27
Figure 3.3	A CNN architecture using LeNet.	34
Figure 4.1	Data description.	52
Figure 4.2	Number of passengers using aircraft.	52
Figure 4.3	Visualization of the differenced data.	53
Figure 4.4	Data prediction using ARIMA (1,1,1) with seasonal order of 12.	54
Figure 4.5	Forecast prediction using ARIMA (1,1,1) model.	54
Figure 4.6	Prediction analysis for future dates.	55
Figure 4.7	Prediction analysis for future dates.	55
Figure 4.8	Prediction using Facebook Prophet Model.	56
Figure 4.9	Trend and yearly prediction.	56
Figure 5.1	Annual revenue in billions of US dollars.	63
Figure 5.2	Shanshan Y et al., collaborative recommendation based on product recommendation.	65

Figure 5.3	Shanshan Y et al., proposed model.	65
Figure 5.4	Observations of Zeng et al. on buyers' online shopping behavior.	67
Figure 5.5	CIB-PA process flow for user requirement prediction and recommendation.	69
Figure 5.6	Model for speech to text using mobile application.	71
Figure 5.7	Comparison of CIB-PA model and existing sales.	73
Figure 5.8	(a) Precision.	73
Figure 5.8	(b) Recall.	74
Figure 6.1	Flow graph of video analytics design using depth intra coding.	82
Figure 6.2	Comparison of complexity, accuracy, and signal to noise ratio.	84
Figure 6.3	Comparison of BJONTEGGARD Delta Bit Rate (BDBR).	84
Figure 6.4	Comparison of time saving.	85
Figure 7.1	Classification of brain tumors in human brain.	89
Figure 7.2	Brain tumor detection system-Image Processing Approach.	91
Figure 7.3	MRI input images.	91
Figure 7.4	Image preprocessing stage.	92
Figure 7.5	Image Segmentation stage.	93
Figure 7.6	Feature extraction stage.	93
Figure 7.7	Brain tumor detection system-Machine Learning Approach.	95
Figure 7.8	MRI brain dataset.	95
Figure 7.9	Image pre-processing stage.	96
Figure 7.10	Image segmentation stage.	96
Figure 7.11	Feature extraction stage.	97
Figure 7.12	Image classification stage.	97
Figure 7.13	Nano-robotic system-A novel approach for automatic detection of brain tumor.	98
Figure 8.1	Our Framework model.	109
Figure 8.2	Confusion Matrix for k-NN.	110
Figure 8.3	Confusion Matrix for Random Forest.	110
Figure 8.4	Confusion Matrix for Neural Networks.	111
Figure 8.5	Test and score for prediction.	111
Figure 8.6	Comparison of different AI approaches in terms of prediction.	111

Figure 9.1	Overall architecture of the fault tolerance deep analysis secured data.	127
Figure 9.2	Implementation architecture of the sensational attacks analyzing fault tolerance ensemble classification using MDNN.	127
Figure 9.3	MDNN architecture.	128
Figure 9.4	The system architecture with the Ensemble classifier model.	132
Figure 9.5	Random forest classification.	140
Figure 9.6	Accuracy comparison of various existing techniques with Proposed Ensemble by applying KDD Cup 99 Dataset	143
Figure 9.7	Precision comparison of various existing techniques with Proposed Ensemble by applying KDD Cup 99 Dataset.	143
Figure 9.8	Area under ROC comparison of various existing techniques with Proposed Ensemble by applying KDD Cup 99 Dataset	144
Figure 9.9	Recall comparison of various existing techniques with Proposed Ensemble by applying KDD Cup 99 Dataset.	145
Figure 9.10	F1-Measure comparison of various existing techniques with Proposed Ensemble by applying KDD Cup 99 Dataset	145
Figure 10.1	Flow chart of novel IoT-based agriculture automation using ML.	153
Figure 10.2	Accuracy comparison.	155
Figure 10.3	Performance comparison.	155
Figure 11.1	Main problems in 5G.	169
Figure 13.1	Architecture of the proposed scheme.	204
Figure 13.2	(a) Request process.	207
Figure 13.2	(b) Message/Update process.	207
Figure 13.3	Process delay.	210
Figure 13.4	Integrity Check Bytes.	211
Figure 13.5	Overhead.	211
Figure 14.1	Architecture diagram of Secured E-Tree MSI Technique.	219
Figure 15.1	Spot Robot.	232
Figure 15.2	Robots service in hotels.	233

Figure 15.3	Percentage of travelers pandemic.	234
Figure 15.4	Threat due to Covid-19	234
Figure 16.1	Schematic diagram of the proposed system.	244
Figure 16.2	Architecture Smart IoT Assistant.	246
Figure 16.3	Internal construction of Smart Assistant	248
Figure 16.4	Build design of Smart Assistant.	250
Figure 16.5	User Interface of the government administration.	251
Figure 16.6	Output viewed by the user.	251
Figure 16.7	A table shows different language output.	251

List of Tables

Table 1.1	Performance evaluation for CNN Model 1	9
Table 1.2	Performance evaluations for CNN Model 2	9
Table 1.3	Performance evaluations for CNN Model 3	10
Table 1.4	Classification results of different CNN Models of OSCC	11
Table 2.1	Speech to text characteristics	20
Table 3.1	Summary of machine learning and deep learning-based CAD systems	35
Table 4.1	Result of Dicky–Fuller test without differencing	52
Table 4.2	Result of Dicky–Fuller test with differencing	53
Table 5.1	Summary of buyer's actions on online products by Zeng et al.	67
Table 5.2	The survey results of the buyer recommendation accuracy	72
Table 6.1	Comparison of parameters	83
Table 9.1	Comparison of the existing and the proposed Ensemble classifier using NSL-KDD dataset	142
Table 13.1	Components and their functions	205
Table 13.2	Comparative Analysis Results	212

List of Contributors

Anbarasi, A., *Research Scholar, Department of computer science & Engineering, Pondicherry University, Puducherry, E-mail: anbarasi.a@gmail.com*

Archana, P., *Sri Manakula Vinayagar Engineering College, Puducherry, India; E-mail: aurchana85@gmail.com*

Arulmurugan, A., *Associate Professor, Department of CSE, Vignan's Foundation for Science, Technology & Research (Deemed to be University), Guntur, Andhra Pradesh, India; E-mail: Arulmurugan1982@gmail.com*

Babukarthik, R. G., *Department of Computer Science & Engineering, Department of Electronic Communication & Engineering, Dayananda Sagar University, Bangalore, India; E-mail: r.g.babukarthik@gmail.com*

Balamurugan, B., *Assistant Professor, Bharathidasan Govt. College for Women, Puducherry, India; E-mail: drbalamurugan@dhtepdy.edu.in*

Balaji, S., *Associate Professor, Department of Information Technology, Sri Manakula Vinayagar Engineering College, Puducherry, India; E-mail: balajisankaralingam@gmail.com*

Fernandez, Terrance Frederick, *Associate Professor, Department of Information Technology, Dhanalakshmi Srinivasan College of Engineering & Technology, Tamil Nadu, India; E-mail: frederick@pec.edu*

Gopi, V. P., *Department of Electronics & Communication Engineering, Sri Manakula Vinayagar Engineering College, Puducherry, India; E-mail: vpgopi2505@gmail.com*

Harikrishnan, M., *Department of Electronics & Communication Engineering, Sri Manakula Vinayagar Engineering College, Puducherry, India; E-mail: harikrishnan@smvec.ac.in*

Jayakumar, B., *Department of Electronics & Communication Engineering, Sri Manakula Vinayagar Engineering College, Puducherry, India; E-mail: jayakumar7120@gmail.com*

Jegadeeswari, S., *Rajiv Gandhi Arts & Science College, Puducherry, India; E-mail: jegadeeswari@dhtepdy.edu.in*

Kaviarasan, R., *Assistant Professor, Department of CSE, RGM College of Engineering and Technology, Nandyal, Andhra Pradesh, India; E-mail: Kaviarasanr64@pec.edu*

Kodati, Sarangam, *Department of CSE, Teegala Krishna Reddy Engineering College, Telangana, India; E-mail: k.sarangam@gmail.com*

Kute, Shruti, *School of Computing Science and Engineering, Vellore Institute of Technology, Chennai, Tamil Nadu, India; E-mail: shrutikute99@gmail.com*

Mahesh, T. R., *Faculty of Engineering and Technology, Department of Computer Science and Engineering, JAIN (Deemed to be University), Bangalore, India; E-mail: t.mahesh@jainuniversity.ac.in*

Malik, Shaveta, *Computer Engineering Department, Terna Engineering College, Navi-Mumbai 600706, Maharashtra, India; E-mail: shavetamalik687@gmail.com*

Manoj Nair, Meghna, *School of Computing Science and Engineering, Vellore Institute of Technology, Chennai, Tamil Nadu, India; E-mail: mnairmeghna@gmail.com*

Manojkiran, K., *Department of Electronics & Communication Engineering, Sri Manakula Vinayagar Engineering College, Puducherry, India; E-mail: mjk1270@gmail.com*

Megala, T., *Acharya Arts and Science College, Puducherry, India; E-mail: divvy.gnsh@gmail.com*

Milan, Aiswariya, *Department of Computer Science & Engineering, Dayananda Sagar University, Bangalore, India; E-mail: aiswariya-cse@dsu.edu.in*

Moganarangan, N., *Associate Professor,Department of Computer Science and Engineering, Rajiv Gandhi College of Engineering and Technology, Puducherry, India; E-mail: rengannsj77@gmail.com*

Mohan Rao, M., *Department of CSE, Ramachandra College of Engineering, Andhra Pradesh, India; E-mail: mohanrao19@gmail.com*

Nelson Kennedy Babu, C., *Professor, Computer Science and Engineering, Saveetha School of Engineering, Saveetha Institute of Medical and Technical Sciences, Chennai, India; E-mail: cnkbabu63@gmail.com*

Palanivel, N., *Associate Professor, Department of Computer Science and Engineering, Manakula Vinayagar Institute of Technology, Puducherry, India; E-mail: npalani76@gmail.com*

Prabavathy, S., *A.P.C. Mahalaxmi College, Thoothukudi, India; E-mail: prabavathy.mk@gmail.com*

Pradeep Reddy, Kumbala, *Department of CSE, CMR Institute of Technology (Autonomous), Hyderabad, Telangana, India; E-mail: pradeep529@gmail.com*

Pradeep, J., *Department of Electronics & Communication Engineering, Sri Manakula Vinayagar Engineering College, Puducherry, India; E-mail: pradeepj@smvec.ac.in*

Pradeep, M., *Assistant Professor, Department of Information Technology, Rajiv Gandhi College of Engineering and Technology, Puducherry, India; E-mail: mpradeepcseb@yahoo.com*

Praveen Kumar, J., *Department of Computer Science and Engineering, Teegala Krishna Reddy Engineering College, Telangana, India; E-mail: praveen@tkrec.ac.in*

Premalatha, S., *Department of Electronics and Communication Engineering, KSR Institute of Engineering and Technology, Tiruchengode, India; E-mail: shineprem1@yahoo.co.in*

Puttagunta, Murali Krishna, *Research Scholar, Department of Computer Science, School of Engineering and Technology, Pondicherry University, Pondicherry, India; E-mail: murali93940@gmail.com*

Raja, G., *Research Scholar, Department of Computer Science & Engineering, Puducherry Engineering College, Puducherry, India; E-mail: raja@pec.edu*

Ravi, G., *Department of CSE, MRCET, Hyderabad, Telangana, India; E-mail: g.raviraja@gmail.com*

Ravi, S., *Associate Professor, Department of Computer Science, School of Engineering and Technology, Pondicherry University, Pondicherry, India; E-mail: sravicite@gmail.com*

Sahoo, Rohit, *Computer Engineering Department, Terna Engineering College, Navi-Mumbai, Maharashtra, India; E-mail: rohitsahoo741@gmail.com*

Salini, P., *Assistant Professor, Department of Computer Science & Engineering, Puducherry Engineering College, Puducherry, India; E-mail: salini@pec.edu*

Saravanan, C., *Faculty of Engineering and Technology, Department of Computer Science and Engineering, JAIN (Deemed to be University), Bangalore, India; E-mail: c.saravanan@jainuniversity.ac.in*

Sindhu Madhuri, G., *Faculty of Engineering and Technology, Department of Computer Science and Engineering, JAIN (Deemed to be University), Bangalore, India; E-mail: g.sindhumadhuri@jainuniversitiy.ac.in*

Sivaramakrishnan, S., *Department of Electronics and Communication Engineering, Dayananda Sagar University, Bangalore, India; Department of Computer Science & Engineering, Department of Electronic Communication & Engineering, Dayananda Sagar University, Bangalore, India; E-mail: sivaramkrish.s@gmail.com*

Tyagi, Amit Kumar, *School of Computing Science and Engineering, Vellore Institute of Technology, Chennai, Tamil Nadu, India; Centre for Advanced Data Science, Vellore Institute of Technology, Chennai, Tamil Nadu, India; E-mail: amitkrtyagi025@gmail.com*

Udaya, D., *Arignar Anna Governments Arts and Science College, Karaikal, India; E-mail: udayadayanadan@gmail.com*

Veeranna, Thotakura, *Department of CSE, Sai Spurthi Institute of Technology, Sathupally, Telangana, India; E-mail: veeru38@gmail.com*

Vijayakumar, K., *Department of Electronics & Communication Engineering, Sri Manakula Vinayagar Engineering College, Puducherry, India; E-mail: msg2world@gmail.com*

Vinay Kumar, K., *Department of Computer Science and Engineering, KITS Warangal, Telangana, India; E-mail: kvk@cse.kits.ac.in*

Vivek, V., *Faculty of Engineering and Technology, Department of Computer Science and Engineering, JAIN (Deemed to be University), Bangalore, India; E-mail: v.vullikanti@jainuniversity.ac.in*

Waris, Saiyed Faiayaz, *Assistant Professor, Department of CSE, Vignan's Foundation for Science, Technology & Research (Deemed to be University), Guntur, Andhra Pradesh, India; E-mail: Saiyed.cse@gmail.com*

List of Abbreviations

AI	Artificial Intelligence
ANN	Artificial Neural Network
API	Application Package Interface
ARIMA	Auto Regression Integrated Moving Average
ATP	Association Time Convention
BQM	Bilinear quadrilateral Mapping
CBC	Cipher Block Chaining
CLI	Command Line Interface
CNN	Convolutional Neural Network
CNN	Convolution Neural Networks
CSV	Comma Separated Value
DDoS	Distributed Denial of Service
DNN	Deep Neural Networks
DoS	Denial of Service
DRNN	Dense Random Neural Network
DSP	Digital Signal Processing
E-R+tree FSQ	E-R+tree Fast Similarity Query
E-tree MSI	Ensemble Tree Metric Space
FFNN	Feedforward Neural Network
FN	False Negative
FP	False Positive
FPLS	Fast Predictive Look-ahead Scheduling approach
Google API	GoogleApplication Programming Interface
GSM	Global System for Mobile Communications
gTTS	Google Text-To-Speech
GUI	Graphical User Interface
HTML	HyperText Markup Language
HTTP	Hypertext Transfer Protocol
ICA	Independent Component Analysis
IDS	Intrusion Detection System
IIES	Improved Image Encryption System

IoT	Internet of Things
LSTM	Long Short-Term Memory
MA	Moving Average
MKQE	Multi-Keyword Query Scheme
ML	Machine Learning
NLP	Natural Language Processing
NS	Normalized Speed
NWC	Normalized Work Capacity
OCR	Optical Character Recognition
OECD	Organization for Economic Co-operation and Development
OTP	One-Time Pad
P-2-P	Peer to Peer
PC	Personal Computer
PDF	Portable Document Format
PETC	Parallel Ensemble Tree Classification
PLC	Payload Capacity
POS	Part of Speech
RNN	Recurrent Neural Network
SMS	Short Message Service
SMT	Statistical Machine Translation
SVD	Singular Value Decomposition
SVM	Support Vector Machine
TCP	Transmission Control Protocol
TN	True Negative
TP	True Positive
TTS	Text-To-Speech
UDP	User Datagram Protocol
URL	Uniform Resource Locator
XML	Extensible Markup Language

1

Classification of Histopathological Variants of Oral Squamous Cell Carcinoma Using Convolutional Neural Networks

P. Archana[1], T. Megala[2], D. Udaya[3], and S. Prabavathy[4]

[1]Sri Manakula Vinayagar Engineering College, Puducherry, India
[2]Acharya Arts and Science College, Puducherry, India
[3]Arignar Anna Governments Arts and Science College, Karaikal, India
[4]A.P.C. Mahalaxmi College, Thoothukudi, India
E-mail: aurchana85@gmail.com; divvy.gnsh@gmail.com; udayadayanadan@gmail.com; prabavathy.mk@gmail.com

Abstract

Oral Cancer is a primary disease which affects the middle age and elderly age group people. When healthy cells undergo mutations, they develop gradually into a mass known as cancer. Cancer can be divided into two types: malignant and benign. The cells in malignant tumors multiply and spread to other parts of the body, whereas benign tumors do not spread. Oral Squamous Cell Carcinoma (OSCC) is a highly prevalent oral cancer that affects more than 90% of the head and neck areas, than other parts of the body. At present, the pathologist finds it difficult to identify the histopathological variants of OSCC. To address this problem, the Computer Aided Diagnosis (CAD) paradigm was developed to assist pathologists in making decisions. However, early detection and prevention of oral cancer is vital, but it is a time-consuming task in medical image processing. Therefore, it is essential to find an effective diagnostic procedure for detecting cancer at the earlier stage. The proposed work aims at developing an effective deep learning model to identify the histopathological variants of OSCC. In the proposed work, convolutional layers and filters size have been altered to attain the accuracy.

To evaluate the performance of the proposed models, a real time dataset is used which shows a highest accuracy of 93.65% to 96.83% compared to other models.

Keywords: Convolutional Neural Network, Oral Squamous Cell Carcinoma, Convolutional Layers.

This work mainly focuses on detecting the affected cells at the earlier stages by using deep learning technique. The proposed Convolutional Neural Network consists of convolutional layers, subsampling layers, and a fully connected layer.

1.1 Introduction

Oral Squamous Cell Carcinoma (OSCC) is the most widespread type of the cancer. Damages in the oral epithelial cells is a collection of numerous genetic mutations displayed in the cells. Approximately, the survival rate of the patients is of five-years, despite the new treatment modalities. The causes of the oral cancer include chewing of tobacco, smoking cigarettes, pipes, and cigars, alcohol consumption, etc. OSCC is found more in males than in females [1]. The occurrence of the oral cancer is mostly found in the fringes of the tongue, floor of the mouth/ventral tongue, and alveolar mucosa/gingiva. These are the most common affected sites. Squamous cell carcinoma is mostly seen in the head as well as the neck region. OSCC can be present as several variants. In Deep Learning, Convolutional Neural Network has been handled for studying the histopathological variants of OSCC.

The main aim of the proposed framework is to classify the OSCC. Figure 1.1 shows the proposed method for classification of histopathological variants of OSCC. The Convolutional Neural Network of the proposed system is designed with a simple stack of multiple convolutional layers proceeded by ReLU activation function followed by max pooling layers. The first Convolutional Neural Network model consists of three convolutional layers. The second and the third Convolutional Neural Network models consist of four and five convolutional layers. By repeating the experiments with different models, the Convolutional Neural Network classifies the histopathological variants of OSCC into Verrucous Squamous Cell Carcinoma, Adenosquamous Cell Carcinoma, Adenoid Squamous Cell Carcinoma, Baseloid Squamous Cell Carcinoma, Papillary Squamous Cell Carcinoma, and Spindle Cell Squamous Carcinoma.

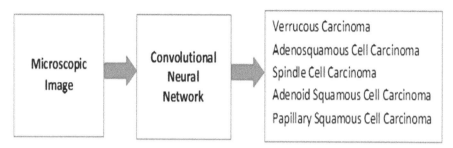

Figure 1.1 Proposed framework of histopathological variants of OSCC.

Every variation of OSCC has a unique histomorphological appearance. The variants include Verrucous Carcinoma, Adenoid Squamous Cell Carcinoma, Spindle Cell Carcinoma, Adenosquamous Cell Carcinoma, and Papillary Squamous Cell Carcinoma. Each variant is discussed below.

- **Verrucous Squamous Cell Carcinoma:** Verrucous Squamous Cell Carcinoma is present as white and wart lesions which are formed as an exophytic growth or well-differentiated squamous cell carcinoma. Oral Verrucous Carcinoma (OVC) is a unique variant of OSCC. OVC is the most uncommon variant of OSCC. The occurrence of Verrucous Carcinoma (VC) of the mouth cavity is between 2% and 16% of all oral cancers [2].
- **Adenoid Squamous Cell Carcinoma:** Adenoid Squamous Cell Carcinoma (ASCC) is an uncommon alternative of squamous cell carcinoma with the characteristics of adenoid pattern of acantholysis. Oral Adenoid Squamous Cell Carcinoma has a unique pseudo vascular morphology [3].
- **Spindle Cell Carcinoma:** Spindle Cell Carcinoma is also referred as the Sarcomatoid Squamous Cell Carcinoma. It is a rare malignancy type found not only in the head but also in the neck region. It mostly originates in larynx, occurs in nasal cavity, oral cavity, esophagus, hypopharynx, etc. Squamous Cell Carcinoma consists of elongated spindle and an epithelial cell that seems like a sarcoma [4].
- **Adenosquamous Cell Carcinoma:** Adenosquamous Cell Carcinoma (ADC) originates in the head and neck and is an invasive different type of squamous cell carcinoma. ADC rarely develops in upper aerodigestive tract, especially in the mouth. The affected areas in the oral cavity are palate, tonsillar pillar areas, and the soft palate [5].

- **Papillary Squamous Cell Carcinoma:** Papillary Squamous Cell Carcinoma (PSCC) is a rare alternative of squamous cell carcinoma. It is mostly prevalent in larynx of the head as well as the neck region. The mouth, oropharynx, sinus, nasal and pharynx are also affected. Within the oral cavity, alveolar ridge, oral mucosa, soft palate, and ventral tongue are the affected areas [6].

1.2 Convolutional Neural Networks

Convolutional Neural Network (CNN) is a collection of Deep Learning which is a subfield of machine learning, is mainly used for image recognition, image classification, video labeling, natural language processing, medical image analysis, etc. The CNN consists of Input Layer, Output layer, and numerous hidden layers. The hidden layers consist of many convolutional layers. ReLU layer is the commonly used activation function followed by the pooling layers, fully connected layers, and normalization layers [7]. Finally, the convolutional layer often uses the back propagation to get more accurate results. Figure 1.2 illustrates the architecture of CNN. Each convolutional layer inside a neural network has the following steps as shown in Figure 1.2.

A data center's general design can be divided into four categories, ranging from Tier I to Tier IV, each with its own set of benefits and drawbacks in terms of power consumption and availability. Most redundant N+1, N+2 or 2N data center architectures are used to address availability and safety concerns, and thus has a significant impact on power usage. A data center, as shown in Figure 1.2, has the following main units.

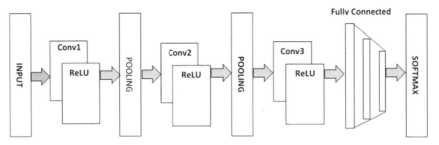

Figure 1.2 Architecture of Convolution Neural Network.

1.2.1 Convolutional Layer

The Convolutional Layer is the structural block of a CNN. The parameters of the layers consist of filters (kernels), a small receptive field, and goes to the depth of the input image. In the progressive pass, each filter is merged to the breadth and length of the input image, calculating the dot product between the filter where the input yields the two-dimensional activation map of that filter.

1.2.2 Pooling Layer

Pooling layer reduces the dimension of the data by merging the output neuron in a single layer as clusters and passes it on to the next layer as a single neuron. For computation, convolutional network pooling includes two types, namely, local pooling and global pooling. Local pooling joins all the small clusters. For example, a 2×2 global pooling poses as a neuron embedded in the convolutional layer [10]. Pooling performs three operations such as Min pooling, Max pooling, and Average pooling. Min pooling takes the minimum value from each group of neurons at the earlier layer; Max pooling adopts the maximum value from every cluster of neurons at the previous layer; and Average pooling accepts the average value from each one of the group of neurons at the preceding layer. The pooling layer gradually alleviates the spatial dimensional representation, decreases the number of parameters, and hence controls the over fitting. The pooling layer works on its own on the input image on each deep slice and reshapes it spatially. The most popular type is the pooling layer with filters of size 2×2 handled with a stride of 2 down sampling at each deep slice in the input by 2 along its [11].

1.2.3 Fully Connected Layers

It uses the principle of multi-layer perceptron (MLP). It connects all the neurons in the single layer to all the neurons in another layer. The fully connected layers classify the images by flattened matrix.

1.2.4 Receptive Field

In the convolutional layer, neurons accept the input from only a limited area of the preceding layer. The insertion of the neuron is called the receptive field. So, the previous layer is the receptive field in the fully connected layer. In the convolutional layer, receptive field takes a trivial place in the preceding layer.

1.2.5 Weights

Each and every neuron in a neural network calculates an output value by handling a particular function to the input values of the preceding level. The function that is used to input the values is estimated by a vector of weights and a bias. The vector of weights and the bias are referred filters that represent a particular feature of the input. A feature of a CNN is that, numerous neurons share the same filter [12].

1.2.6 ReLU Layer

The full form of ReLU is the Rectified Linear Unit which is employed to the non-saturating activation function which is defined as

$$f(x) = max\,(0, x) \qquad (1.1)$$

It removes the negative values from the activation function by placing them to zero.

1.2.7 Softmax Layer

The last layer is the softmax layer. The softmax function particularizes a discrete probability allocation for K classes, represented by following Equation (1.2)

$$\sum_{k=1}^{K} P_K. \qquad (1.2)$$

The last layers of CNN are fully connected, and the final level handles softmax role to its input, so that we obtain probabilities for every single image. The network parameters are trained using back-propagation algorithm as in view of the usual neural networks.

1.2.8 Dropout

The term dropout refers to the dropping up of (hidden or visible) units randomly during the training. During the testing time, it is difficult to handle the overfitting because large networks take a lot of time when joining the predictions of different large neural networks. It reduces overfitting. When a unit is dropped out, the neuron fails to donate to the forward-pass. The dropout neuron gives the probability of 0.5 for each hidden neurons and output to be zero.

1.2.9 Steps Involved in Convolutional Neural Network

The basic steps involved in the process of classification of OSCC using convolutional neural network is given below.

Step 1: Provide input image inside convolutional layer.
Step 2: Choose parameters, apply filters along with strides, padding if necessary.
Step 3: Perform convolution on the image and use ReLU activation to the matrix.
Step 4: Render pooling to decrease dimensionality size.
Step 5: Contribute as many convolutional layers until contented.
Step 6: Flatten the outcome and contribute into a fully connected layer (FC Layer).
Step 7: Outproduce the class handling an activation function.

1.3 Proposed Convolutional Neural Network

The CNN structure of the proposed system is designed with a simple stack of multiple convolutional layers proceeded by ReLU activation function followed by max pooling layers. Figure 1.3 shows the proposed framework of CNN where the three convolutional models are designed with different filter and parameter size.

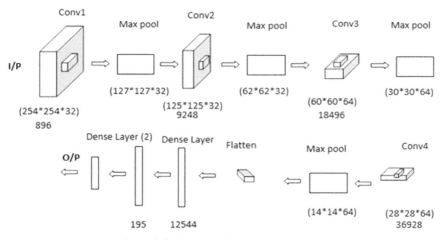

Figure 1.3 Proposed frameworks of CNN.

In the first CNN model, three convolutional layers proceeded by three max pooling layers have been used. The image is resized into $254 \times 254 \times 3$. In the First Convolutional Layer, 32 filters have been convolved to the input image and it yields 896 parameters; Maximum pooling and ReLU is used for the initial convolutional layer.

In the second Convolutional Layer, the same 32 filters, have been utilized which yields 9248 parameters. In the Third Convolutional Layer, 64 filters have been used, they yield 18,496 parameters. Maximum pooling and ReLU have been utilized for all the convolutional Layers. Finally, these are flattened, 57,600 feature dimensions are obtained and it is fed into the Dense Layer. For testing the input image of size $254 \times 254 \times 3$, microscopic images are given for prediction. The prediction of classes is computed based on Keras model prediction analysis. The prediction matrix is given to the five classes namely Verrucous Carcinoma, Spindle Cell Carcinoma, Papillary Squamous Cell Carcinoma, Adenosquamous Cell Carcinoma, and Adenoid Squamous Cell Carcinoma.

1.3.1 Performance Evaluation for CNN Models

The performance of the proposed three-CNN model is calculated with performance metrics namely Precision, Recall, F-Score, and Accuracy, based on the confusion matrix on the predicted value. In order to evaluate the performance of the proposed work, a real-time dataset was collected from Rajah Muthiah Dental College and Hospital (RMDC & H). The images were collected with $20\times$ magnifications of Hematoxylin and Eosin staining images. A total of 350 images were collected, in which 225 were used for training, 125 were used for testing. Here five-fold cross validation have been used. First Convolutional Layer, 32 filters have been convolved to the input image and it yields 896 parameters, Maximum pooling and ReLU are used for the initial convolutional layer. The performance evaluation of model 1 is presented in Table 1.1 below.

In the second CNN model, four convolutional layers preceded by four max pooling layers have ben used. The image is resized into $254 \times 254 \times 3$. For training and testing, the image is resized into $254 \times 254 \times 3$. In the First Convolutional Layer, 32 filters have been convolved to the input image and it yields 896 parameters. Maximum pooling and ReLU are used for the initial convolutional layer. In the second Convolutional Layer, the same 32 filters, have been used which yield 9248 parameters. In the Third Convolutional Layer, 64 filters have been used; it yields 18,496 parameters. In the fourth

Table 1.1 Performance evaluation for CNN Model 1

Cell Classification	Precision (in %)	Recall (in %)	F-Score (in %)	Accuracy (in %)
Verrucous Squamous Cell Carcinoma	72.00	90.00	80.00	92.90
Adenoid Squamous Cell Carcinoma	76.00	83.00	79.00	92.00
Spindle Cell Carcinoma	88.00	88.01	88.19	95.20
AdenoSquamous Cell Carcinoma	84.00	72.00	78.00	90.40
Papillary Squamous Cell Carcinoma	88.00	79.21	83.00	92.80

Convolutional Layer, the same 64 filters have been used which yields 36,928 parameters. Maximum pooling and ReLU have been utilized for the all the Convolutional Layers. Finally, it is flattened, 12,544 feature dimensions are obtained and these are fed into the Dense Layer. For testing the input image of size $254 \times 254 \times 3$, microscopic images are given for prediction. The prediction of classes is computed based on Keras model prediction analysis. The prediction matrix is rendered to the five classes, namely, Verrucous Carcinoma, Spindle Cell Carcinoma, Papillary Squamous Cell Carcinoma, Adenosquamous Cell Carcinoma, and Adenoid Squamous Cell Carcinoma. The performance of CNN model 2 is presented in Table 1.2.

In the third CNN model, five convolutional layers proceeded by five max pooling layers have been used. The image is resized into $254 \times 254 \times 3$. For training and testing the image is resized into $254 \times 254 \times 3$. In the First

Table 1.2 Performance evaluations for CNN Model 2

Cell Classification	Precision (in %)	Recall (in %)	F-Score (in %)	Accuracy (in %)
Verrucous Squamous Cell Carcinoma	88.00	96.00	92.00	96.77
Adenoid Squamous Cell Carcinoma	88.00	91.00	89.00	95.97
Spindle Cell Carcinoma	76.00	90.01	83.19	93.55
AdenoSquamous Cell Carcinoma	80.00	71.00	75.00	89.52
Papillary Squamous Cell carcinoma	96.00	83.21	89.00	95.16

Table 1.3 Performance evaluations for CNN Model 3

Cell Classification	Precision (in %)	Recall (in %)	F-Score (in %)	Accuracy (in %)
Verrucous Squamous Cell Carcinoma	80.00	99.86	89.00	96.03
Adenoid Squamous Cell Carcinoma	84.00	91.00	88.00	95.24
Spindle Cell Carcinoma	88.00	92.01	90.19	96.03
AdenoSquamous Cell Carcinoma	92.00	80.00	86.00	93.65
Papillary Squamous Cell carcinoma	99.85	86.00	93.00	96.83

Convolutional Layer, 32 filters have been convolved to the input image and this yields 896 parameters. Maximum pooling and ReLU are used for the initial convolutional layer. In the second Convolutional Layer, the same 32 filters, have been used which yield 9248 parameters. In the Third Convolutional Layer, 64 filters have been utilized; they yield 18,496 parameters. In the fourth Convolutional Layer, the same 64 filters have been used which yield 36,928 parameters. In the fifth Convolutional Layer, the same 64 filters have been used which yield 147,520 parameters. Maximum pooling and ReLU have been used for all the Convolutional Layers. Finally, they are flattened, 2304 feature dimensions are obtained and these are fed into the Dense Layer. For testing the input image of size $254 \times 254 \times 3$, microscopic images are given for prediction. The prediction of classes is computed, based on Keras model prediction analysis. The prediction matrix is given to the five classes, namely, Verrucous Carcinoma, Spindle Cell Carcinoma, Papillary Squamous Cell Carcinoma, Adenosquamous Cell Carcinoma, and Adenoid Squamous Cell Carcinoma. The performance of CNN model 3 is presented in Table 1.3.

1.3.2 Comparative Result Analysis

The result attained by the different CNN models with different layers and different filters is analyzed. From the results, it is analyzed that model 3 gives the best accuracy of 96.83% when compared to model 1 and model 2, where models 1 and 2 have an accuracy less than 96%. The graphical representation of comparison of all three models in identifying and classifying the variants of OSCC is represented in Figure 1.4.

1.3 Proposed Convolutional Neural Network

Table 1.4 Classification results of different CNN Models of OSCC

Histopathological Variants of OSCC	CNN Model 1 (in %)	CNN Model 2 (in %)	CNN Model 3 (in %)
Verrucous Squamous Cell Carcinoma	92.90	96.77	96.03
Adenoid Squamous Cell Carcinoma	92.00	95.97	95.24
Spmdle Cell Carcinoma	95.20	93.55	96.03
AdenoSquamous Cell Carcinoma	90.40	89.52	93.65
Papillary Squamous Cell carcinoma	92.80	95.16	96.83

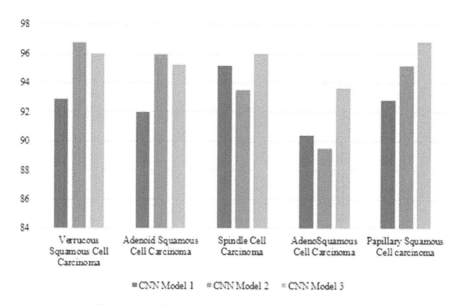

Figure 1.4 Classification results of different CNN models.

Based on the classification results between the three models, CNN model 2 shows the highest accuracy of 96.77% for Verrucous Squamous Cell Carcinoma, CNN model 2 shows the highest accuracy of 95.97% for Adenoid Squamous Cell Carcinoma, CNN model 3 shows an accuracy of 96.03% for Spindle Cell Carcinoma, 93.65% for AdenoSquamous Cell Carcinoma, and the highest accuracy of 96.83% for papillary Squamous Cell Carcinoma. Based on the result, it is evident that compared with CNN model 1 and CNN model 2, CNN model 3 outperforms the other two models.

1.4 Conclusion

This chapter aims at developing an effective deep learning model to identify the histopathological variants of Oral Squamous Cell Carcinoma. The work mainly focuses on detecting the affected cells at the earlier stages by using deep learning technique. A convolutional neural network model with different filter and parameter size is proposed. The proposed three models classify the Histopathological variants of OSCC into Verrucous Squamous Cell Carcinoma, Adenosquamous Cell Carcinoma, Adenoid Squamous Cell Carcinoma, Papillary Squamous Cell Carcinoma, and Spindle Cell Carcinoma based on morphology of the cell. A real time dataset has been used to evaluate the performance of the proposed models. Based on accuracy, precision, recall and F-Score of the three models, model 3 shows highest accuracy of 93.65% to 96.83% when compared to model 1 and model 2. The outcome of the proposed work concludes that convolutional neural networks give satisfactory results in classifying the histopathological variants of OSCC.

Acknowledgments

We thank all the hospital staffs and the advisory board of Rajah Muthiah Dental College and Hospital (RMDC & H) for their valuable contribution to this project work. We thank our mentors and co-researchers for their valuable comments and advice throughout the project.

References

[1] Pires, Fabio Ramoa, et al. "Oral squamous cell carcinoma: clinicopathological features from 346 cases from a single oral pathology service during an 8-year period." Journal of Applied Oral Science 21.5 (2013): 460–467.

[2] Franklyn, Joshua, et al. "Oral verrucous carcinoma: ten year experience from a tertiary care hospital in India." *Indian journal of medical and paediatric oncology: official journal of Indian Society of Medical & Paediatric Oncology* 38.4 (2017): 452.

[3] Hassan, M. I., et al. "Correlation of Lymphovascular Density with Histological Prognostic Parameters in Gastric Carcinoma." *Journal of Histopathology and Cytopathology* 4.1 (2020): 12–22.

[4] Matsukuma, Susumu, et al. "Esophageal adenosquamous carcinoma mimicking acantholytic squamous cell carcinoma." *Oncology letters* 14.4 (2017): 4918–4922.

[5] Sravya, Taneeru, et al. "Oral adenosquamous carcinoma: Report of a rare entity with a special insight on its histochemistry." *Journal of oral and maxillofacial pathology: JOMFP* 20.3 (2016): 548.

[6] Fitzpatrick, Sarah G., et al. "Papillary variant of squamous cell carcinoma arising on the gingiva: 61 cases reported from within a larger series of gingival squamous cell carcinoma." *Head and neck pathology* 7.4 (2013): 320–326.

[7] Aubreville, Marc, et al. "Automatic classification of cancerous tissue in laserendomicroscopy images of the oral cavity using deep learning." *Scientific reports* 7.1 (2017): 1–10.

[8] Fu, Qiuyun, et al. "A deep learning algorithm for detection of oral cavity squamous cell carcinoma from photographic images: A retrospective study." *EClinicalMedicine* 27 (2020): 100558.

[9] Halicek, Martin, et al. "Hyperspectral imaging of head and neck squamous cell carcinoma for cancer margin detection in surgical specimens from 102 patients using deep learning." *Cancers* 11.9 (2019): 1367.

[10] Sakellariou GK, Pearson T, Lightfoot AP, Nye GA, Wells N, Giakoumaki II, Vasilaki A, Griffiths RD, Jackson MJ, McArdle A. Mitochondrial ROS regulate oxidative damage and mitophagy but not age-related muscle fiber atrophy. Sci Rep. 2016 Sep. 29; 6:33944. doi: 10.1038/srep33944. PMID: 27681159; PMCID: PMC5041117.

[11] Springenberg, J.T., Dosovitskiy, A., Brox, T. and Riedmiller, M., 2014. Striving for simplicity: The all convolutional net. *arXiv preprint arXiv:1412.6806*.

[12] Sermanet, P., Eigen, D., Zhang, X., Mathieu, M., Fergus, R. and LeCun, Y., 2013. Overfeat: Integrated recognition, localization and detection using convolutional networks. *arXiv preprint arXiv:1312.6229*.

2

Voice Recognition Using Natural Language Processing

J. Pradeep, K. Vijayakumar, and M. Harikrishnan

Department of Electronics & Communication Engineering, Sri Manakula Vinayagar Engineering College, Puducherry, India
E-mail: pradeepj@smvec.ac.in; msg2world@gmail.com; harikrishnan@smvec.ac.in

Abstract

Natural Language Processing (NLP) is a method used to describe the relationship between machines with human language. This chapter describes an online voice-based test for the physically disabled applicants. To extract the data from the candidates through learning, a new process called voice recognition using NLP is implemented and through which speech is transformed into text format. The machine will store a text document detailing the actual response. The text format is correlated with text documents using the Spacy algorithm, and both responses do not need to be the same word-by-word. Mark will be distributed on the grounds of the similarity between the two sentences.

Keywords: NLP Algorithm, Speech to Text Conversion, Voice Recognition.

2.1 Introduction

Understanding and representing the human language is a difficult task, because it is a discrete system and it has ambiguity due to the complexity of the language's expression and interpretation [14]. There is a need to implement a new technology that eradicates the process' difficulty. Natural Language Processing (NLP) is a branch of artificial intelligence that deals with human and machine interactions with the use of natural language [13]. It is a computer program's capacity to recognize the human language spoken and to perform various tasks, such as answering questions, reading text, evaluating emotions, marking expression, and generally understanding speech. Data analysis using linguistic tools is the possibility of creating a high-performance NLP system [1]. NLP also verifies that the resulting sentence is spoken after the recognition of expression [2]. NLP also made it possible for industries to process more accurate and automated processing of speech and text [3, 19].

Voice recognition is a hardware tool or a program capable of interpreting the human voice and converting them into machine-readable format. In simple terms, it is the process through which a system can comprehend natural language [4, 15]. Speech-based systems are much friendlier, comfortable, and more accessible than the type-in based method [11]. The use of NLP is not only limited to chat or text interfaces but has also arisen as a leading speech recognition tool [16]. Complex speech patterns can also be recognized at a decent speed and can be converted into its text format with the help of NLP [5]. Semantic identification is also essential for an effective speech to text conversion that can be accomplished by NLP [6]. The integration of speech technology and Internet technology makes voice communication a new area of research [7].

Persons with benchmark disabilities such as blindness, locomotors disability, or a cerebral palsy are provided with the facility of using a scribe or a reader on the production of a certificate to the effect that the person concerned has physical limitation to write, and a scribe is essential to write the examination on their behalf. On such conditions, the candidate has the discretion of opting for a scribe at his own cost or may request the examination body for the same, which includes meeting the scribe two days before the examination for verifying whether the scribe is suitable or not.

The next step involves the evaluation process, which includes three stages. In the first stage, an examiner evaluates all the exam papers and allocates mark manually. The next step involves a first level moderator who re-evaluates 10% to 20% of the answer sheets evaluated by the initial examiner. The third stage involves a second level moderator who re-evaluates 5% of the answer sheets evaluated by the first level moderator. If there is any difference in the valuation of the paper, changes are made. To overcome these complications, a voice recognition system using NLP is introduced.

2.2 Proposed System

In this section, the proposed recognition system is described. Voice recognition system upon which Google API acts as platform. It is inbuilt for pre-processing, segmentation, feature extraction, classification, and recognition. The schematic diagram of the proposed system is shown in Figure 2.1.

2.2.1 Automatic Speech Recognition

It is a kind of technology that provides human beings to use their voice to speak with the machines like computer interface, which resemble a normal human, begins conversation. Real-time streaming or pre-recorded audio supports the audio input from an application's microphone and it can be streamed or sent from a particular pre-recorded audio file.

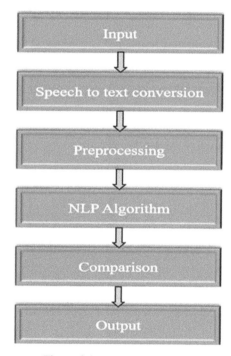

Figure 2.1 Proposed method.

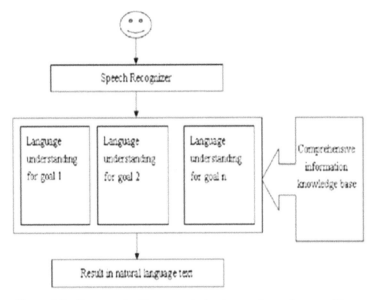

Figure 2.2 Framework of language technology and speech recognition.

2.2.2 Auto-detect Language

This application detects the language spoken by the person by using Google API, automatically. The General Framework of language technology and speech recognition is shown in the Figure 2.2.

2.2.3 Syntactic Analysis

The Syntactic analysis premiere checks the grammatical reliability of each word. For grammatical information, researchers use part-of-speech (POS) usually. But here, we did not use it because POS needs lots of time and memory and what's more; sentences after Speech Recognition are different from normal correct text in that it may be wrong sentences that have illegal POS usage.

2.2.4 Semantic Analysis

The Semantic analysis premiere checks the capability of a word to illuminate the meaning of the sentence. Normally, there are more than one word in a sentence. Some of them are used mainly to construct the sentence, which are functional words, such as imperative words, pronouns, quantifier words, auxiliary words, etc. Other words are mainly used to illuminate the meaning

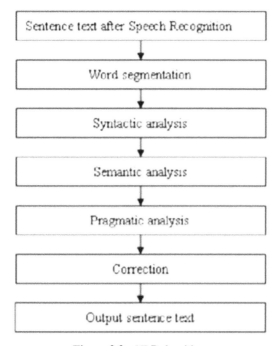

Figure 2.3 NLP algorithm.

of a sentence. Integration of the two kinds of words could make a good sentence.

2.2.5 Pragmatic Analysis

The Pragmatic analysis premiere checks the harmony degree between a word and its surrounding environment for a given goal.

In this proposed method, we use the advanced NLP algorithm to find the meaning of the sentence by using the abovementioned techniques and comparing it with the output obtained from the speech to text conversion and it approves the existing method. The NLP algorithm schematic diagram is shown in Figure 2.3.

2.3 Experimental Results

The recognition system has been implemented in the computer using Google API. The audio is recorded live and speech to text conversion method is used. Then the text is compared with the text documents of the original answer paper. In this method of speech to text conversion using Google API, the features involved are powered machine learning algorithm, recognition of many languages, automatic identification of spoken language, speech adaptation, noise robustness, and model selection. Using the NLP algorithm like Spacy, the text from the speech and the text in the documents are compared. This proposed method is implemented with many inputs with different voices and the output is obtained. The advantage of this method is that all the speaking words by the candidate will be able to appear on the display screen. Its highly accurate based on the speech recognition tools.

Table 2.1 indications the Speech to text characteristics for the performance of the speech to text conversion. Table 2.1 indications the Speech to text characteristics for the performance of the speech to text conversion.

The input is given to the computer through means of voice, the speech to conversion process is done, and text is displayed in the PC. Once the candidate confirms the text sentence then it is sent to the file that holds all the original answers in the text document.

Table 2.1 Speech to text characteristics

Characteristics	Remark
Open Source	Yes
Features	Good
Training	Easy
Performance	Good
Usage	Fair
Accuracy	99%

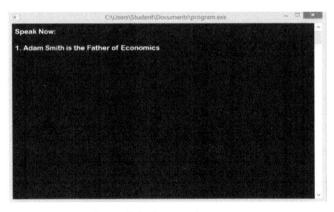

Figure 2.4 Input to the system.

2.3 Experimental Results 21

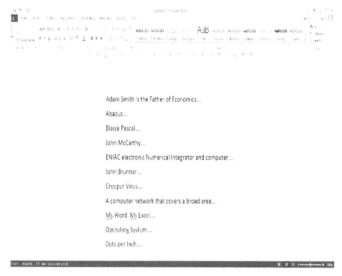

Figure 2.5 Text document.

The output from the speech to text conversion is given as the input to text documents that compares the text sentences with the answer in the file and evaluates the marks. The Output Screen Text document is shown in Figure 2.5

Once the marks are evaluated, then the system renders a response message to the candidate to proceed with the further question. The Output Screen for the given input is shown in Figure 2.6

Figure 2.6 Output screen.

2.4 Conclusion

A simple voice recognition system for physically challenged students using a new concept of voice recognition using NLP algorithm is proposed. Experimental results show that the voice recognition method with NLP method yield good recognition of 98.75% with 50 test cases and it produces 96.45% with 200 test cases. The result obtained and presented in this paper shows that the voice recognition system using NLP method is better than the manual method that exists. This proposed method with API is suitable for many applications like voice-based ordering system in hotel and online interview with computers. The other applications, which are discussed in this paper, are work in progress.

References

[1] A. Gupta, N. Patel and S. Khan, "Automatic speech recognition technique for voice command," 2014 International Conference on Science Engineering and Management Research (ICSEMR), Chennai, 2014, pp. 1–5.

[2] R. Shanmugapriya and S. RajaMohammed, "Speech recognition open source tools for the semantic identification of the sentence," 2014 International Conference on Green Computing Communication and Electrical Engineering (ICGCCEE), Coimbatore, 2014, pp. 1–3.

[3] S. Sitaram, S. Palkar, Y. Chen, A. Parlikar and A. W. Black, "Bootstrapping Text-to-Speech for speech processing in languages without an orthography," 2013 IEEE International Conference on Acoustics, Speech and Signal Processing, Vancouver, BC, 2013, pp. 7992–7996.

[4] Li HongLian, Wang ChunHua and Yuan BaoZong, "Using context knowledge to improve the accuracy of recognition in voice browsing," 6th International Conference on Signal Processing, 2002, Beijing, China, 2002, pp. 544–547, vol. 1.

[5] K. M. Shivakumar, V. V. Jain and P. K. Priya, "A study on impact of language model in improving the accuracy of speech to text conversion system," 2017 International Conference on Communication and Signal Processing (ICCSP), Chennai, 2017, pp. 1148–1151.

[6] O. Turk and L. M. Arslan, "Speech recognition methods for speech therapy," Proceedings of the IEEE 12th Signal Processing and Communications Applications Conference, 2004, Kusadasi, Turkey, 2004, pp. 410–413.

[7] L. Li, Y. Zhou, L. Zhang and Y. Zhong, "Enhancing Speech Based Information Service with Natural Language Processing," 2006 4th IEEE International Conference on Industrial Informatics, Singapore, 2006, pp. 1057–1062.

[8] S. Karthick, R. J. Victor, S. Manikandan and B. Goswami, "Professional chat application based on natural language processing," 2018 IEEE International Conference on Current Trends in Advanced Computing (ICCTAC), Bangalore, 2018, pp. 1–4.

[9] T. Nasukawa, "Text analysis and knowledge mining," 2009 Eighth International Symposium on Natural Language Processing, Bangkok, 2009, pp. 1–2.

[10] P. Withanage, T. Liyanage, N. Deeyakaduwe, E. Dias and S. Thelijjagoda, "Road Navigation System Using Automatic Speech Recognition (ASR) And Natural Language Processing (NLP)," 2018 IEEE Region 10 Humanitarian Technology Conference (R10-HTC), Malambe, Sri Lanka, 2018, pp. 1–6.

[11] H. Isahara, "Resource-based Natural Language Processing," 2007 International Conference on Natural Language Processing and Knowledge Engineering, Beijing, 2007, pp. 11–12.

[12] M. Uma, V. Sneha, G. Sneha, J. Bhuvana and B. Bharathi, "Formation of SQL from Natural Language Query using NLP," 2019 International Conference on Computational Intelligence in Data Science (ICCIDS), Chennai, India, 2019, pp. 1–5.

[13] N. Tyson and V. C. Matula, "Improved lsi-based natural language call routing using speech recognition confidence scores," Second IEEE International Conference on Computational Cybernetics, 2004, pp. 409–413.

[14] D. Black, E. J. Rapos and M. Stephan, "Voice-Driven Modeling: Software Modeling Using Automated Speech Recognition," 2019.

[15] G. Muhammad, M. Alsulaiman, A. Mahmood and Z. Ali, "Automatic voice disorder classification using vowel formants," 2011 IEEE International Conference on Multimedia and Expo, Barcelona, 2011, pp. 1–6. ACM/IEEE 22nd International Conference on Model Driven Engineering Languages and Systems Companion (MODELS-C), Munich, Germany, 2019, pp. 252–258.

3

Detection of Tuberculosis Using Computer-Aided Diagnosis System

Murali Krishna Puttagunta[1], S. Ravi[2,*], A. Anbarasi[3], and Nelson Kennedy Babu C[4]

[1]Research Scholar, Department of Computer Science, School of Engineering and Technology, Pondicherry University, Pondicherry, India
[2]Associate Professor, Department of Computer Science, School of Engineering and Technology, Pondicherry University, Pondicherry, India
[3]Research Scholar, Department of Computer Science, School of Engineering and Technology, Pondicherry University, India
[4]Professor, Computer Science and Engineering, Saveetha School of Engineering, Saveetha Institute of Medical and Technical Sciences, Chennai, India
E-mail: murali93940@gmail.com; sravicite@gmail.com; anbarasi.a@gmail.com; cnkbabu63@gmail.com
*Corresponding Author

Abstract

A systematic analysis of artificial intelligence-based computer-aided diagnostic (CAD) systems for analyzing chest X-rays for diagnosing pulmonary tuberculosis (TB) is carried out in this paper. Tuberculosis (TB) is a transmissible disease that is one of the top 10 causes of death globally. In TB-affected countries, there is a strong need to improve care and screening. The four major areas in the literature on CAD system analysis of chest radiographs are: (i) pre-processing techniques, (ii) image segmentation, (iii) feature extraction, and (iv) classification. A detailed survey of the development of various phases of CAD programs for the diagnosis of TB is presented in this paper. The development of CAD systems aids in the early detection of TB.

Keywords: Chest X-ray Imaging, Pulmonary Tuberculosis, Computer-Aided Diagnosis, Machine Learning, Automated Screening.

3.1 Introduction

Since the 1970s, researchers have been developing computer-aided diagnostic (CAD) systems to help in the identification of tuberculosis (TB) during the screening phase [1]. CAD systems have an automated detection system which can be used to assist radiologists in remote areas [2]. Recent advances in artificial intelligence (AI) research and techniques have resulted in major changes in automatic machine image recognition [3]. CAD systems uses artificial intelligence to identify abnormalities in radiological images, alleviating the shortage of radiologists, especially in developing countries [4, 5]. For the identification of breast cancer [7], Alzheimer's disease [6], skin lesions [8], neurological disorders [10], and lung cancer [9], CAD systems are widely used. The two most commonly applied AI methods for developing CAD systems capable of analyzing Chest X-ray (CXRs) are Machine Learning (ML) [11] and Deep Learning (DL) [12].

With the advancement of science and technology, as well as the emergence of medical imaging modality, the analysis and interpretation of medical image data has become more complicated for radiologists [13]. When bacteriological tests fail to provide a satisfactory result, radiology becomes critical [14]. TB is caused by the bacterium *Mycobacterium tuberculosis*. Cough, weight loss, night sweats, fever, exhaustion, anorexia, and chest pain are all common symptoms of active TB [15]. The World Health Organization (WHO) recommends high-quality CXR images and laboratory-based diagnostic testing for early TB diagnosis and to minimize the TB detection gap [16].

CXR is an important method for routine screening of patients with TB symptoms and identifying the need of microscopic sputum examination, according to WHO. The majority of people with TB symptoms are assessed using a CAD method to analyze their CXR. By checking CXR images, referring to people for a bacteriological test is shown in Figure 3.1.

Medical imaging acts as a powerful tool to gain knowledge of the anatomy and function of an organ, other than the diagnosis of diseases. Image-processing approaches on medical imaging combined with machine learning techniques lead to CAD systems [18]. While TB disease is most commonly associated with the lungs, it can affect any organ system

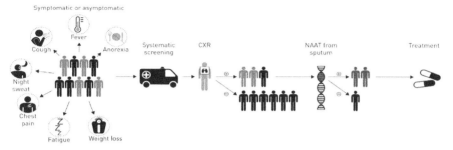

Figure 3.1 Chest radiography (CXR) with computer-aided detection (CAD) used as a triage test [17].

in the body. Chest radiography (CXR) is an effective method for monitoring and screening pulmonary TB [19]. The initial scan for an unidentified cough is a CXR posteroanterior (PA) picture of the lung. Pneumothorax, pulmonary consolidation, pleural effusion, nodules, invasion, atelectasis, emphysema, and cardiac hypertrophy are all diseases that CAD systems can detect in CXR [20]. The anatomical structure of the chest is depicted in Figure 3.2.

The ML-based CAD system has the following phases: (i) preprocessing, (ii) segmentation, (iii) feature extraction, and (iv) classification. We describe all the phases of the CAD system in the following sections. In these systems, first, the image is preprocessed to improve image quality, then the region of interest (ROI) is segmented and the features are extracted from the ROI using handcraft methods. Extracted feature is vector fed to the classifier where CXR is classed as normal or abnormal [21].

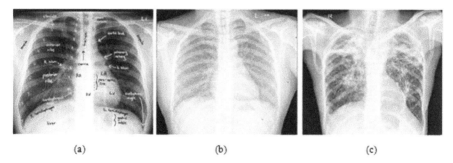

Figure 3.2 (a) Chest anatomy, (b) A healthy chest X-ray image, (c) Tuberculosis with multiple cavitation's CXR images.

3.2 Pre-Processing

A CAD system usually applies a series of preprocessing steps to an input image [22]. For CAD systems, unprocessed images are not suitable for processing because of poor contrast, sensor noise, and superimposed anatomical structures. The quality of preprocessing steps further affects the subsequent systems. The main aim of preprocessing is to enhance image quality and suppress different anatomical structures so that nodules and opacities become easily detectable. Different preprocessing steps of an X-ray image are image enhancement, anatomical noise suppression [23], bone suppression, edge sharping, filtering [24], and image subtraction techniques.

Histogram equalization is the common image enhancement technique, which adjusts image intensity values, frequent intensity values of the image pixels are spread out to enhance the contrast of the image. Different histogram equalization techniques are constrained with local histogram equalization [25, 26], gray-level histogram equalization [27], and adaptive histogram equalization [28] and have been employed for increasing the quality of the CXR image. Other enhancement techniques such as the piecewise linear model, energy normalization technique, and transformations based on wavelet [29, 30] were proposed in the literature. Bone suppression is a preprocessing technique used to remove ribs and clavicle structures. The authors have proposed a rib subtraction technique based on the local PCA-based shape model of the rib profiles [31]. The effect of rib suppression on the classification performance of a CAD system for TB detection is evaluated.

3.3 Segmentation

Segmentation means the image is divided into related regions. Depending on the level of the feature extraction, segmentation can be methodically classified into four categories: (i) rule-based, (ii) pixel classification-based, (iii) deformable model-based, and (iv) hybrid methods.

3.3.1 Rule-Based Algorithm

The rule-based algorithm consists of a certain series of steps and adjustable parameters to segment the lung region. The authors proposed a CAD system for automated lung segmentation using Global and Local gray level Threshold-based approach [32] in digitized posteroanterior CXR. In [33], the authors proposed the edge detection with the first derivatives of the image to extract the ROI besides applying the iterate contour smoothing algorithm to smooth the lung boundary in lung segmentation.

3.3.2 Pixel Classification

Pixel classification is a classical segmentation algorithm that works based on pixel intensity. Each pixel is classified to be inside or outside the lung fields by the classifier. For this, it uses a large amount of annotated data for training classifiers. Mcnitty-gray et al. [34] proposed a lung segmentation method using a neural network-based classifier for the classification of anatomical classes based on local features of the CXR image. Vittitoe et al. [35] proposed the Markov random field model for the segmentation of the lungs by using textural and spatial information of the lungs.

3.3.3 Deformable Models

Deformable models used for lung segmentation are divided into two categories, namely, parametric and geometric models. Active Shape Model (ASM) and Active Appearance Model (AAM) are parametric models applied for the segmentation of the lung region [36]. ASM was first proposed by T.F. Cootes [37] by developing a hand-annotated set of images. With the marking of landmark points based on the training set of images, new shapes suitable to the test image are created. T.F. Cootes [38] further changed the original algorithm to find optimal landmark points to fit the shape variability of the lungs. Xu et al. [39] proposed a Gradient vector flow ASM to segment the lung field. Active shape and appearance models have limitations because of the shape of rib cage edges, initial model assumptions, and internal parameters. The authors used the ASM technique to segment the lung fields [40].

Graph cut and level set methods are geometric models. The graph cut model minimizes the objective function for the segmentation of the lungs [41]. The objective function is modeled mathematically. If a binary vector $f = \{f_1, f_2, f_3 \ldots f_p, \ldots f_n\}$ whose components f_p corresponds to (f_g) foreground and (b_g) background labels assignments to pixel $p \in P$, where P is the set of pixels in the X-ray image, and N is the number of pixels. The objective function $E(f)$ was defined as:

$$E(f) = E_d(f) + E_s(f) + E_m(f) \qquad (3.1)$$

where E_d, E_s and E_m are the data, smoothness, and lung model properties of the X-ray image respectively. By using a fast implementation of the min-cut/max-flow algorithm, the objective function for the optimal configuration of f is minimized. The minimum objective function separates the input CXR image into the foreground (lung)/background configuration.

3.3.4 Hybrid Methods

Hybrid methods perform better by combining several techniques. The proposed hybrid method combined the rule-based algorithm with a pixel classification algorithm [42]. Three hybrid models were proposed in combining ASM, AAM, and pixel classification schemes by majority voting [43]. In [44], the authors proposed a non-rigid registration-driven robust lung segmentation method for CXR images. This method was used for lung segmentation in TB detection process [45–48].

3.4 Feature Extraction

Textural features are developed for explaining honeycomb-structured lung tissue from an X-ray image. For computing texture features from the X-ray image, different procedures are used. Texture features can be extracted from the gray level distribution of intensity values of an image. Statistical feature extraction methods are used for texture analysis. In literature, the shape and texture descriptors used in TB detection are described below.

3.4.1 Histogram Features

Intensity histogram is a first-order statistical feature, which provides information about the distribution of gray levels in the image. The probability distribution of intensity values in the image is $P(i)$, where

$$P(i) = \frac{Number\ of\ pixels\ with\ gray\ level\ i}{number\ of\ pixels\ in\ the\ image} = \frac{N(i)}{M} \qquad (3.2)$$

The four first-order histogram statistical features Mean, Variance, Skewness, and Kurtosis are obtained from Intensity histogram [49, 50].

$$Mean: \mu = \sum_{i=0}^{N-1} iP(i) \qquad (3.3)$$

$$Variance: \sigma = \sum_{i=0}^{N-1} (i-\mu)^2 P(i) \qquad (3.4)$$

$$\text{Skewness}: s = \sum_{i=0}^{N-1}(i-\mu)^3 P(i) \tag{3.5}$$

3.4.2 Shape Descriptor Histogram

Shape descriptor for Lung region is defined as

$$SD = \tan^{-1}\left(\frac{\lambda_1}{\lambda_2}\right) \tag{3.6}$$

where λ_1 and λ_2 are the Eigenvalues of the Hessian matrix.

3.4.3 Curvature Descriptor

Curvature descriptor describes curvature properties of lung image and is expressed as

$$CD = \tan^{-1}\left(\frac{\sqrt{\lambda_1^2 + \lambda_2^2}}{1 + I(x,y)}\right) \tag{3.7}$$

with $0 \leq CD \leq \frac{\pi}{2}$, where $I(x,y)$ denotes pixel intensity.

3.4.4 Local Binary Pattern (LBP)

LBP was originally proposed by Ojala et al. in 1996 [51]. LBP combines structural methods and statistical methods. The LBP method represents a binary pattern for each pixel in an image. The LBP codes are calculated as the gray level of the central pixel can be considered as a threshold and compared with the eighth-pixel values of the neighborhood [52]. The pixel threshold value is multiplied by the weights and summed up to get a single value. An LBP code is 0 or 1. The codes are ranging from 0 to $2^L - 1$. It is formulated as the following Equation (3.8) and (3.9):

$$LBP_{LR}(I_c) = \sum_{L=0}^{L-1} S(x) 2^L \tag{3.8}$$

$$S(x) = \begin{cases} 1 & if\, x \geq 0 \\ 0 & otherwise \end{cases} \tag{3.9}$$

where $x = I_N - I_c$, I_N = Intensity of neighborhood pixel within a circle, I_C = Intensity of central pixel within a circle, L = Neighboring pixels are on a circle, R = Radius of circle. Authors have used LBP feature for TB detection from x-ray images [46, 47,53–55].

3.4.5 Histogram of Gradients

Histogram of Gradients (HOG) can provide the edge direction by extracting the gradient and orientation of the edge. HOG features were used for local object detection in a CXR image [46–48,55–57]. This approach has been used to split a CXR image into small cells, measure the HOGs for each cell and provide a descriptor. The descriptor generation is composed of four major steps: (i) Computing the gradient image in x and y, (ii) Computing gradient histogram, (iii) Normalizing across blocks, and (iv) Flattening into a feature vector.

For each cell, the gradient was calculated. Gradients are small change in x and y directions. x and y derivatives of an image I_x and I_y are computed by using the convolution operation $I_x = I * D_x$ and $I_y = I * D_y$.

where $D_x = [-1\ 0\ 1]$, $D_y = [-10\ 1]^T$. Gradient magnitude and orientation are calculated by the following Equations (3.10) and (3.11).

$$|G| = I_x^2 + I_y^2 \tag{3.10}$$

$$\theta = \tan^{-1}\frac{I_x}{I_y} \tag{3.11}$$

3.4.6 Gabor Features

Gabor wavelet transform $G_{mn}(x, y)$ [56] for an image $f(x, y)$ is

$$G_{mn}(x, y) = \int f(x, y) g_{mn}^*(x - x_1, y - y_1) dx_1 dy_1 \tag{3.12}$$

g_{mn}^* Represents complex conjugate of $g(x, y)$.

$$g_{mn}(x, y) = a(x', y')\ 0 \leq m \leq M - 1, 0 \leq n \leq N - 1 \tag{3.13}$$

where $x' = a^{-m}(x\cos\theta + y\sin\theta)$, $y' = a^{-m}(x\sin\theta + y\cos\theta)$ and $\theta = \frac{n\pi}{k}k$, is the total number of orientations.

The μ_{mn} and σ_{mn} are the magnitudes of Gabor wavelet transform coefficients and are expressed as:

$$\text{Mean } \mu_{mn} = \int\int |G_{mn}(x, y)|\ dxdy \tag{3.14}$$

$$\text{Standard deviation } \sigma_{mn} = \sqrt{\int\int (|G_{mn}(x, y)| - \mu_{mn})^2\ dxdy} \tag{3.15}$$

These are used for representing regions for classification and for retrieving images.

3.5 Classification

Support Vector Machine (SVM) was proposed based on the statistical learning theory [59]. SVM aimed to identify a hyperplane that separates the two classes of data points with a maximal separation margin that serves as the boundary for decision [60]. SVMs are another class of machine learning algorithms and have received considerable recognition in the TB detection literature [45, 47, 54, 56, 57,61–63]. TB detection is a two-class classification problem.

Given a training dataset has m training examples with labeled Instance pair $\{(x_i, y_i), i = 1, 2, \ldots m\}$ $x_i \in R^n$ and $y_i \in \{-1, +1\}$ [64]. The goal is to construct decision function (classifier) f(x) that accurately classifies the labels of unseen data and minimizes the classification error. So the decision function is defined as

$$f(x) = wx + b \tag{3.16}$$

where w is a vector that is perpendicular to the hyperplane, and b specifies the offset value. For each training example, x_i the function $f(x_i) \geq 0$ for $y_i = +1$ and $f(x_i) < 0$ for $y_i = -1$. The hyperplane or decision boundary to separate the classes is $f(x) = wx + b = 0$. For a training dataset, there are many hyperplanes to separate the two classes. The main aim of the SVM classifier was to separate the two classes of data points in which hyperplane has a maximum margin between the two classes. Finding the optimal hyperplane is the same as the optimization problem of the quadratic function. This optimization problem is handled by the Lagrange function L and Lagrange multiplier α_i.

$$L(w, b, \alpha) = \frac{1}{2}\|w\|^2 - \sum_{i=1}^{m} \alpha_i(y_i((x_i \cdot w) + b) - 1) \tag{3.17}$$

By using Karush–Kuhn–Tucker conditions, ultimately the decision function written as

$$f(x) = sgn\left(\sum_{i=1}^{m} y_i \alpha_i (x \cdot x_i) + b\right) \tag{3.18}$$

For linearly non-separable case, it is necessary to introduce an additional scale variable ξ_i ($\xi \geq 0$)

$$y_i((w_i \cdot x_i) + b) \geq 1 - \xi_i i = 1, 2 \ldots . m \tag{3.19}$$

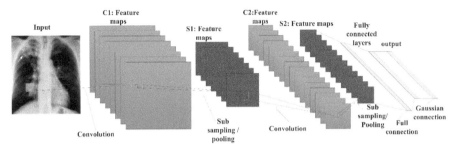

Figure 3.3 A CNN architecture using LeNet.

The vector w determines the optimal hyperplane

$$\min \quad \phi(w, \xi) = \frac{1}{2}\|w\|^2 + C \sum_{i=1}^{m} \xi_i \qquad (3.20)$$

where C is a user-defined parameter that controls the trade-off between the margin and the slack penalty. For non-linear classification, Kernel functions are used [65, 66].

Deep learning is a new and rapidly emerging area that provides excellent solutions to a variety of CAD problems. In the field of radiology, DL has quickly become the technique of choice [41]. Deep Convolutional Neural Networks (CNNs) have been instrumental in the extraction of features for TB disease identification and classification of CXR images as normal or abnormal. For TB detection, several different CNN variants have been suggested, including LeNet [67], AlexNet [68], GoogleNet [69], VGGNet [70], DenseNet [71], ResNet [72], and R-CNN. Convolutional layers, subsampling/pooling layers, and completely linked layers are typical components of CNN models. Figure 3.3 shows basic CNN architecture for TB classification.

3.6 Discussion

This paper reviews and summarizes research papers on TB detection from CXR images using CAD systems published between 1999 and 2020. Artificial Neural Networks (ANNs) are used to construct CAD systems because they perform better in Pattern Recognition problems. When it comes to prediction of outcomes, Neural Networks outperform traditional statistical approaches. A comparison sheet is created in Table 3.1, summarizing machine learning and deep learning-based CAD systems

3.6 Discussion

Table 3.1 Summary of machine learning and deep learning-based CAD systems

Author & Reference Number	Year	Segmentation Technique	Feature Selection	Classification Technique	Dataset/ No. of CXR	Accuracy (%)
El-solh et.al. [73]	1999	--	Demographic variables, constitutional symptoms, radiographic findings	General Regression Neural Network (GRNN)	--	92.3
Hariharan et al. [74]	2000	Fuzzy Kohonen Neural Network	--	--	--	--
Mhod.Noor et.al [75]	2002	--	Continuous wavelet transform (CWT)	Clustering	100	--
Arzhaeva et al. [76]	2009		Gaussian derivatives Central movements	MVDB	461	AUC:0.83
Shen et al. [77]	2010	Automated the initial contour placement	GICOV and circularity measure	Bayesian classification	56	82.35
Jaeger et al. [61]	2012	Intensity mask The lung model masks the Log Gabor mask	Shape and texture descriptors	SVM classifier	JSRT, MC	83.12
Jaeger et al. [45]	2014	Graph cut-based lung segmentation	Object detection features and CBIR-based image features	Support vector machine (SVM)	JSRT, MC, CH	82
Melendez et al. [78]	2016	--	Feature Ranking	Multiple learner fusion	Zambian	84
Ding et al. [46]	2017	Graph cut-based lung boundary segmentation	Frangi, HOG, LBP GIST	Local (SVM) Global (SVM+RBF) classifier fusion	CH, IN, Kenya	90(CH) 96(IN) 73(Kenya)
Jeyavathana et al. [47]	2017	Graph cut-based segmentation	LHTGF feature (LBP, HOG, Tamura, and Gabor filter)	SVM classifier	CH	89
Govindarajan [79]	2019	ACS-based distance regularized level set	Bag of Features (BoF) approach with Speeded-Up Robust Feature (SURF)	Multilayer Perceptron (MLP) classifier.	MC	85.9
Hwang et al. [80]	2016	--	--	AlexNet with transfer learning	KIT	96
Liu et al. [81]	2018	--	--	CNN	4701	85.68
Devnath et al. [82]	2018	--	--	CNN	MC, CH, IN (1078)	87.29
Ghorakavi [53]	2019	--	LBP, Harr features	Deep-ConvNet Resnet-18	800	58.35 65.77
S. J. Heo et al., [83]	2019	--	--	D-CNN (Image CNN + Demographic variables	2000	AUC:0.921
Y. Ban et al. [84]	2020	--	--	CNN	TBX11K	89.7
Bram Van Ginniken et al. [40]	2002	ASM	Texture features	k-NN	388	82

(Continued)

Table 3.1 (Continued)

Rajaraman et al. [48]	2018	Graph cut-based lung segmentation	HOG/GIST/SURF CNN	SVM Logistic regression	CH, MC, IN, Kenya	0.93 (CH) 0.875 (MC) 0.96 (IN)
Vajda et al. [55]	2018	Adaptive segmentation	CLD, Gabor, GLCM, CF, IH, HOG, LBP	Neural network-based classifier	MC, CH	97
Hasan et al. [65]	2017	Golden standard mask	Spatial pyramid of SURF	SVM classifier with sigmoid kernel	MC	72.5
Andayani et al. [85]	2019	Active contour segmentation	Invariant moment	PNN	--	96

MC, Montgomery County; CH, Shenzhen Hospital, China; JSRT, Japanese Society of Radiology; IN, Indian collection New Delhi; KIT, Korean Institute of Tuberculosis; TBX11K, Tuberculosis 11200 X-ray images; PHOG, Pyramid Histogram of Oriented Gradients; LHFC, LBP and HOG based feature classification; KLD, Kullback–Leibler divergence; GICOV, Gradient Inverse Coefficient Of Variation; kNN, k-nearest neighbor; LoG, Laplacian of Gaussian; PNN, Probability Neural Network; LDA, Linear Discriminant Classifier; MVDB, Multi-Valued Dissimilarity Based classifier; AUC, Area Under Curve.

The authors developed a General Regression Neural Network (GRNN) using clinical and radiographic information to predict pulmonary TB [73]. In [74], an algorithm is proposed for the enhancement of CXR images of TB patients using a fuzzy-based enhancement technique. For segmentation, the authors use a Fuzzy Kohonen Neural Network. In [75], continuous wavelet transforms are used on CXR images to extract the features. Wavelet transforms divide the CXR image into low- and high-frequency components. For the identification of TB in a CXR image, the authors used a statistical analysis clustering method on these features.

Arzhaeva et al. [76] proposed a first CAD system focused on the classification of weakly labeled images using Multi-valued Dissimilarity-Based classification (MVDB). Shen et al. [77] proposed a hybrid knowledge-guided detection technique for TB cavity detection. Firstly, automatic initialization of suspected cavity's detection is done. Then, these cavities are segmented out by Bayesian classifier using Gradient Inverse Coefficient of Variation (GICOV) and M circularity measure features. This method has an accuracy of 82.35%.

Jaeger et al. [61] proposed an inexpensive TB detection system method using combined lung masks on posteroanterior CXR images. The authors first segmented the lung field by combining intensity mask, log Gabor mask, and statistical lung model mask and then extracted shape, texture, and curvature features. The authors also use SVM as a classifier to identify normal and abnormal CXR images and obtained AUC of 83.12%. Jaeger et al. [45] described an automated screening method for CXR to detect TB. Graph cut segmentation method was used to extract the lung region and then shape and texture features were computed so that the binary classifier can classify two data sets of CXR images into normal and abnormal.

Melendez et al. [78] presented an X-ray based CAD system combined with clinical information for TB screening. For this, the authors introduced ML-based combination framework, which was evaluated using a patient record consisting of CAD score computed from X-ray image and 12 clinical features. The authors showed that the performance of combined CAD and clinical information based on TB detection was better than the individual strategies. Ding et al. [46] developed a local–global classifier fusion method for abnormality detection in CXR images.

Jeyavathana et al. [47] introduced LHTGF (LBP, HOG, Tamura, and Gabor filter) feature extraction method of TB detection on CXRs. The authors combined Gabor features with LBP, HOG, and Tamura features to get LHTGF feature extraction method which uses a multi-class SVM classifier for classification of X-rays. Using the LHTGF feature extraction method authors got 93.4% accuracy in Montgomery dataset and 89% accuracy in the China data set. In this work, [79] authors have attempted to extract the features from CXR images using the Bag of Features (BoF) approach with Speeded-Up Robust Feature (SURF) keypoint descriptor. Multilayer Perceptron (MLP) classifier is implemented for categorizing the images as normal and TB CXR images from BoF.

In remote and resource-constrained areas, CAD systems are critical for enhancing the diagnostic effectiveness of CXRs for TB screening. Our research shows that to accurately detect TB on CXR images, a CAD framework based on DL algorithms is needed. The method of automated feature extraction is the key benefit of using deep learning over other methods. In the identification of TB, CNNs are a promising method for CAD systems.

3.7 Conclusion

This paper explored a variety of image processing techniques for TB detection in CXR images, including preprocessing, segmentation, and feature extraction techniques. CAD systems are designed to detect and locate tuberculosis (TB) through a thorough examination of CXR images. However, the World Health Organization (WHO) has not approved any of the CAD systems for TB detection. For the WHO to approve the use of CAD in TB detection, more evidence is required [86]. This paper's systematic analysis of TB detection emphasizes the need for further research into CAD programs that use CXR for TB detection. AI-based CAD systems are expected to play a critical role in TB detection [87].

References

[1] B. Van Ginneken, B. M. Ter Haar Romeny, and M. A. Viergever, "Computer-aided diagnosis in chest radiography: A survey," *IEEE Trans. Med. Imaging*, vol. 20, no. 12, pp. 1228–1241, 2001, doi: 10.1109/42.974918.

[2] R. Hooda, A. Mittal, and S. Sofat, "Tuberculosis Detection from Chest Radiographs?: A Comprehensive Survey on Computer-Aided Diagnosis Techniques," *Curr. Med. Imaging Rev.*, vol. 14, no. 4, p. 506, 2018, doi: 10.2174/1573405613666171115154119.

[3] K. Suzuki, *Machine Learning in Computer-Aided Diagnosis: Medical Imaging Intelligence and Analysis: Medical Imaging Intelligence and Analysis*. 2012.

[4] C. Qin et al., "Computer-aided detection in chest radiography based on artificial intelligence: A survey," *Biomed. Eng. Online*, vol. 17, no. 1, pp. 1–23, Aug. 2018, doi: 10.1186/s12938-018-0544-y.

[5] U. Raghavendra, U. R. Acharya, and H. Adeli, "Artificial Intelligence Techniques for Automated Diagnosis of Neurological Disorders," *Eur. Neurol.*, vol. 43210, pp. 1–24, 2019, doi: 10.1159/000504292.

[6] M. Adel et al., "Alzheimer's Disease Computer_Aided Diagnosis on Positron Emission Tomography Brain Images using Image Processing Techniques," in *Computer Methods and Programs in Biomedical Signal and Image Processing*, 2019, p. 13.

[7] A. Osareh and B. Shadgar, "Machine learning techniques to diagnose breast cancer," *2010 5th Int. Symp. Heal. Informatics Bioinformatics, HIBIT 2010*, pp. 114–120, 2010, doi: 10.1109/HIBIT.2010.5478895.

[8] I. Bakkouri and K. Afdel, "Computer-aided diagnosis (CAD) system based on multi-layer feature fusion network for skin lesion recognition in dermoscopy images," *Multimed. Tools Appl.*, vol. 79, pp. 20483–20518, 2019, doi: 10.1007/s11042-019-07988-1.

[9] M. Firmino, G. Angelo, H. Morais, M. R. Dantas, and R. Valentim, "Computer-aided detection (CADe) and diagnosis (CADx) system for lung cancer with likelihood of malignancy," *Biomed. Eng. Online*, vol. 15, no. 1, pp. 1–17, 2016, doi: 10.1186/s12938-015-0120-7.

[10] A. A. A. Valliani, D. Ranti, and E. K. Oermann, "Deep Learning and Neurology: A Systematic Review," *Neurol. Ther.*, vol. 8, no. 2, pp. 351–365, 2019, doi: 10.1007/s40120-019-00153-8.

[11] Q. Huang, F. Zhang, and X. Li, "Machine Learning in Ultrasound Computer-Aided Diagnostic Systems: A Survey," *Biomed Res. Int.*, vol. 2018, 2018, doi: 10.1155/2018/5137904.

[12] G. Chartrand *et al.*, "Deep learning: A primer for radiologists," *Radiographics*, vol. 37, no. 7, pp. 2113–2131, 2017, doi: 10.1148/rg.2017170077.

[13] M. Wernick, Y. Yang, J. Brankov, G. Yourganov, and S. Strother, "Machine learning in medical imaging," *IEEE Signal Process. Mag.*, vol. 27, no. 4, pp. 25–38, 2010, doi: 10.1109/MSP.2010.936730.

[14] A. C. Nachiappan *et al.*, "Pulmonary Tuberculosis?: Role of Radiology in Diagnosis and Management," *Radiographics*, vol. 37, no. 1, pp. 52–72, 2017.

[15] "INDEX-TB Guidelines on extra-pulmonary tuberculosis for India," 2016.

[16] "CHEST RADIOGRAPHY IN TUBERCULOSIS DETECTION," *WHO*, 2016.

[17] F. A. Khan *et al.*, "Computer-aided reading of tuberculosis chest radiography?: moving the research agenda forward to inform policy," 2017, doi: 10.1183/13993003.00953-2017.

[18] Y. H. Chan, Y. Z. Zeng, H. C. Wu, M. C. Wu, and H. M. Sun, "Effective pneumothorax detection for chest X-ray images using local binary pattern and support vector machine," *J. Healthc. Eng.*, vol. 2018, 2018, doi: 10.1155/2018/2908517.

[19] A. S. Bhalla, A. Goyal, R. Guleria, and A. K. Gupta, "Chest tuberculosis?: Radiological review and imaging recommendations," *Indian J. Radiol. Imaging*, vol. 25, no. 3, p. 213, 2015, doi: 10.4103/0971-3026.161431.

[20] M. Cicero *et al.*, "Training and Validating a Deep Convolutional Neural Network for Computer-Aided Detection and Classification of Abnormalities on Frontal Chest Radiographs," *Invest. Radiol.*, vol. 52, no. 5, pp. 281–287, 2017, doi: 10.1097/RLI.0000000000000341.

[21] R. Hooda, A. Mittal, and S. Sofat, "Automated TB classification using ensemble of deep architectures," *Multimed. Tools Appl.*, vol. 78, no. 22, pp. 31515–31532, 2019, doi: 10.1007/s11042-019-07984-5.

[22] S. Jaeger *et al.*, "Automatic screening for tuberculosis in chest radiographs?: a survey," vol. 3, no. 2, pp. 89–99, 2013, doi: 10.3978/j.issn.2223-4292.2013.04.03.

[23] O. Tischenko, C. Hoeschen, and E. Buhr, "Reduction of anatomical noise in medical X-ray images," *Radiat. Prot. Dosimetry*, vol. 114, no. 1–3, pp. 69–74, 2005, doi: 10.1093/rpd/nch518.

[24] B. Y. M. K. and H. K. Kwan, "Improved Lung Nodule Visualization on Chest Radiographs using Digital Filtering and Contrast Enhancement,"

World Acad. Sci. Eng. Technol. Vol5, vol. 4, no. 12, pp. 1633–1636, 2011, [Online]. Available: http://waset.org/publications/6635/improved-lung-nodule-visualization-on-chest-radiographs-using-digital-filtering-and-contrast-enhancement.

[25] D. C. Chang and W. R. Wu, "Image contrast enhancement based on a local standard deviation model," *IEEE Nucl. Sci. Symp. Med. Imaging Conf.*, vol. 3, no. 4, pp. 1826–1830, 1996, doi: 10.1109/nssmic.1996.587984.

[26] K. Koonsanit, S. Thongvigitmanee, N. Pongnapang, and P. Thajchayapong, "Image enhancement on digital X-ray images using N-CLAHE," *BMEiCON 2017 - 10th Biomed. Eng. Int. Conf.*, vol. 2017-Janua, no. August 2017, pp. 1–4, 2017, doi: 10.1109/BMEiCON.2017.8229130.

[27] ravin C. Mcadams HP, johnson GA, suddarth SA, "histogram -directed processing of digital chest images."

[28] G. A. Johnson, "Regionally Adaptive Histogram Equalization of the Chest," vol. MI, no. 1, pp. 1–7, 1987.

[29] T. Matozaki, A. Tanishita, and T. Ikeguchi, "Image enhancement of chest radiography using wavelet analysis," *Annu. Int. Conf. IEEE Eng. Med. Biol. - Proc.*, vol. 3, pp. 1109–1110, 1996, doi: 10.1109/iembs.1996.652731.

[30] X. Xu, Y. Wang, G. Yang, and Y. Hu, "Image enhancement method based on fractional wavelet transform," *2016 IEEE Int. Conf. Signal Image Process. ICSIP 2016*, pp. 194–197, 2017, doi: 10.1109/SIPROCESS.2016.7888251.

[31] L. E. Hogeweg, C. Mol, P. A. de Jong, and B. van Ginneken, "Rib suppression in chest radiographs to improve classification of textural abnormalities," *Med. Imaging 2010 Comput. Diagnosis*, vol. 7624, p. 76240Y, 2010, doi: 10.1117/12.844409.

[32] S. G. Armato, M. L. Giger, K. Ashizawa, and H. MacMahon, "Automated lung segmentation in digital lateral chest radiographs," *Med. Phys.*, vol. 25, no. 8, pp. 1507–1520, 1998, doi: 10.1118/1.598331.

[33] L. Li, Y. Zheng, M. Kallergi, and R. A. Clark, "Improved method for automatic identification of lung regions on chest radiographs," *Acad. Radiol.*, vol. 8, no. 7, pp. 629–638, 2001, doi: 10.1016/S1076-6332(03)80688-8.

[34] M. F. McNitt-Gray, J. W. Sayre, and H. K. Huang, "Feature Selection in the Pattern Classification Problem of Digital Chest Radiograph Segmentation," *IEEE Trans. Med. Imaging*, vol. 14, no. 3, pp. 537–547, 1995, doi: 10.1109/42.414619.

[35] N. F. Vittitoe, R. Vargas-Voracek, and C. E. Floyd, "Identification of lung regions in chest radiographs using Markov random field modeling," *Med. Phys.*, vol. 25, no. 6, pp. 976–985, 1998, doi: 10.1118/1.598405.
[36] S. Candemir, "A review on lung boundary detection in chest X-rays," *Int. J. Comput. Assist. Radiol. Surg.*, vol. 14, no. 4, pp. 563–576, 2019, doi: 10.1007/s11548-019-01917-1.
[37] T. Cootes, C. Taylor, D. Cooper, and J. Graham, "Active Shape Models-Their training and Applications," *Computer Vision and Image Understanding*, vol. 61, no. 1. pp. 38–59, 1995.
[38] T. F. Cootes, G. J. Edwards, and C. J. Taylor, "Active appearance models," *Lect. Notes Comput. Sci. (including Subser. Lect. Notes Artif. Intell. Lect. Notes Bioinformatics)*, vol. 1407, no. 6, pp. 484–498, 1998, doi: 10.1007/BFb0054760.
[39] T. Xu, M. Mandal, R. Long, I. Cheng, and A. Basu, "An edge-region force guided active shape approach for automatic lung field detection in chest radiographs," *Comput. Med. Imaging Graph.*, vol. 36, no. 6, pp. 452–463, 2012, doi: 10.1016/j.compmedimag.2012.04.005.
[40] B. Van Ginneken, S. Katsuragawa, M. Bart, H. Romeny, K. Doi, and M. A. Viergever, "Automatic Detection of Abnormalities in Chest Radiographs Using Local Texture Analysis," vol. 21, no. 2, pp. 139–149, 2002.
[41] S. Candemir, S. Jaeger, K. Palaniappan, S. Antani, and G. Thoma, "Graph Cut Based Automatic Lung Boundary Detection in Chest Radiographs," *IEEE Healthc. Technol. Conf. Transl. Eng. Heal. Med.*, no. January, pp. 31–34, 2012.
[42] B. Van Ginneken and B. M. Ter Haar Romeny, "Automatic segmentation of lung fields in chest radiographs," *Med. Phys.*, vol. 27, no. 10, pp. 2445–2455, 2000, doi: 10.1118/1.1312192.
[43] B. Van Ginneken, M. B. Stegmann, and M. Loog, "Segmentation of anatomical structures in chest radiographs using supervised methods?: a comparative study on a public database Revised version," vol. 10, pp. 19–40, 2004, doi: 10.1016/j.media.2005.02.002.
[44] S. Candemir *et al.*, "Lung segmentation in chest radiographs using anatomical atlases with nonrigid registration," *IEEE Trans. Med. Imaging*, vol. 33, no. 2, pp. 577–590, 2014, doi: 10.1109/TMI.2013.2290491.
[45] S. Jaeger *et al.*, "Automatic tuberculosis screening using chest radiographs," *IEEE Trans. Med. Imaging*, vol. 33, no. 2, pp. 1–13, 2013, doi: 10.1109/TMI.2013.2284099.

[46] M. Ding *et al.*, "Local-global classifier fusion for screening chest radiographs," *Med. Imaging 2017 Imaging Informatics Heal. Res. Appl.*, vol. 10138, no. February, p. 101380A, 2017, doi: 10.1117/12.2252459.
[47] R. B. Jeyavathana, R. Balasubramanian, and A. Pandian, "An Efficient Feature Extraction Method for Tuberculosis detection using Chest Radiographs," vol. 12, no. 2, pp. 227–240, 2017.
[48] S. Rajaraman *et al.*, "A novel stacked generalization of models for improved TB detection in chest radiographs," *Proc. Annu. Int. Conf. IEEE Eng. Med. Biol. Soc. EMBS*, vol. 2018-July, no. July, pp. 718–721, 2018, doi: 10.1109/EMBC.2018.8512337.
[49] I. Gabriella, K. S. A, and S. A. W, "Early Detection of Tuberculosis using Chest X-Ray (CXR) with Computer-Aided Diagnosis," *2018 2nd Int. Conf. Biomed. Eng.*, pp. 76–79, 2018.
[50] J. H. Tan, U. R. Acharya, C. Tan, K. T. Abraham, and C. M. Lim, "Computer-Assisted Diagnosis of Tuberculosis?: A First Order Statistical Approach to Chest Radiograph," pp. 2751–2759, 2012, doi: 10.1007/s10916-011-9751-9.
[51] T. Ojala, M. Pietikäinen, and D. Harwood, "A comparative study of texture measures with classification based on feature distributions," *Pattern Recognit.*, vol. 29, no. 1, pp. 51–59, 1996, doi: 10.1016/0031-3203(95)00067-4.
[52] J. M. Carrillo-De-Gea, G. García-Mateos, J. L. Fernández-Alemán, and J. L. Hernández-Hernández, "A Computer-Aided Detection System for Digital Chest Radiographs," *J. Healthc. Eng.*, vol. 2016, 2016, doi: 10.1155/2016/8208923.
[53] Ghorakavi and R. Srivatsav, "TB Net: Pulmonary Tuberculosis Diagnosing System using Deep Neural Networks," *arXiv Prepr. arXiv1902.08897.*, 2019, [Online]. Available: http://arxiv.org/abs/1902.08897.
[54] T. Xu, I. Cheng, R. Long, and M. Mandal, "Novel coarse-to-fine dual scale technique for tuberculosis cavity detection in chest radiographs," pp. 1–18, 2013.
[55] S. Vajda *et al.*, "Feature Selection for Automatic Tuberculosis Screening in Frontal Chest Radiographs," *J. Med. Syst.*, vol. 42, no. 8, 2018, doi: 10.1007/s10916-018-0991-9.
[56] A. Chauhan, D. Chauhan, and C. Rout, "Role of Gist and PHOG Features in Computer-Aided Diagnosis of Tuberculosis without Segmentation," *PLoS One*, vol. 9, no. 11, p. e112980, 2014, doi: 10.1371/journal.pone.0112980.

[57] A. Fatima, M. U. Akram, M. Akhtar, and I. Shafique, "Detection of tuberculosis using hybrid features from chest radiographs," *Eighth Int. Conf. Graph. Image Process. (ICGIP 2016)*, vol. 10225, no. Icgip 2016, p. 102252B, 2017, doi: 10.1117/12.2266795.

[58] A. Zotin, Y. Hamad, K. Simonov, and M. Kurako, "Lung boundary detection for chest X-ray images classification based on GLCM and probabilistic neural networks," *Procedia Comput. Sci.*, vol. 159, pp. 1439–1448, 2019, doi: 10.1016/j.procs.2019.09.314.

[59] J. Nalepa and M. Kawulok, "Selecting training sets for support vector machines: a review," *Artif. Intell. Rev.*, vol. 52, no. 2, pp. 857–900, 2019, doi: 10.1007/s10462-017-9611-1.

[60] S. Saxena and M. Gyanchandani, "Machine Learning Methods for Computer-Aided Breast Cancer Diagnosis Using Histopathology?: A Narrative Review," *J. Med. Imaging Radiat. Sci.*, pp. 1–12, 2019, doi: 10.1016/j.jmir.2019.11.001.

[61] S. Jaeger, A. Karargyris, S. Antani, and G. Thoma, "Detecting Tuberculosis in Radiographs Using Combined Lung Masks," pp. 4978–4981, 2012.

[62] A. Karargyris *et al.*, "Combination of texture and shape features to detect pulmonary abnormalities in digital chest X-rays," *Int. J. Comput. Assist. Radiol. Surg.*, vol. 11, no. 1, pp. 99–106, 2016, doi: 10.1007/s11548-015-1242-x.

[63] K. M. Sundaram and C. S. Ravichandran, "An adaptive region growing algorithm with support vector machine classifier for Tuberculosis cavity identification," *Am. J. Appl. Sci.*, vol. 10, no. 12, pp. 1616–1628, 2013, doi: 10.3844/ajassp.2013.1616.1628.

[64] Chen Junli and Jiao Licheng, "Classification mechanism of support vector machines," pp. 1556–1559, 2002, doi: 10.1109/icosp.2000.893396.

[65] F. Hasan, O. Alfadhli, A. A. Mand, S. Sayeed, and K. S. Sim, "Classification of Tuberculosis with SURF Spatial Pyramid Features," *Int. Conf. Robot. Autom. Sci.*, 2017, doi: 10.1109/ICORAS.2017.8308044.

[66] I. El-Naqa, Y. Yang, M. N. Wernick, N. P. Galatsanos, and R. M. Nishikawa, "A support vector machine approach for detection of microcalcifications," *IEEE Trans. Med. Imaging*, vol. 21, no. 12, pp. 1552–1563, 2002, doi: 10.1109/TMI.2002.806569.

[67] Y. LeCun, L. L. Bottou, Y. Bengio, P. Ha, and P. Haffner, "Gradient-based learning applied to document recognition," *Proc. IEEE*, vol. 86, no. 11, pp. 2278–2323, 1998, doi: 10.1109/5.726791.

[68] A. Krizhevsky, I. Sutskever, and G. E. Hinton, "ImageNet Classification with Deep Convolutional Neural Networks," in *the 25th International Conference on Neural Information Processing Systems*, 2012, pp. 1097–1105, doi: 10.1145/3065386.

[69] C. Szegedy, S. Reed, P. Sermanet, V. Vanhoucke, and A. Rabinovich, "Going deeper with convolutions," pp. 1–12.

[70] K. Simonyan and A. Zisserman, "Very deep convolutional networks for large-scale image recognition," in *3rd International Conference on Learning Representations, ICLR 2015*, 2015, pp. 1–14.

[71] G. Huang, Z. Liu, L. Van Der Maaten, and K. Q. Weinberger, "Densely connected convolutional networks," *Proc. - 30th IEEE Conf. Comput. Vis. Pattern Recognition, CVPR 2017*, vol. 2017-Janua, pp. 2261–2269, 2017, doi: 10.1109/CVPR.2017.243.

[72] K. He, X. Zhang, S. Ren, and J. Sun, "Deep residual learning for image recognition," *Proc. IEEE Comput. Soc. Conf. Comput. Vis. Pattern Recognit.*, vol. 2016-Decem, pp. 770–778, 2016, doi: 10.1109/CVPR.2016.90.

[73] A. A. El-Solh, C. Bin Hsiao, S. Goodnough, J. Serghani, and B. J. B. Grant, "Predicting active pulmonary tuberculosis using an artificial neural network," *Chest*, vol. 116, no. 4, pp. 968–973, 1999, doi: 10.1378/chest.116.4.968.

[74] S. Hariharan, A. K. Ray, and M. K. Ghosh, "Algorithm for the enhancement of chest X-ray images of tuberculosis patients," *Proc. IEEE Int. Conf. Ind. Technol.*, vol. 2, pp. 107–112, 2000, doi: 10.1109/icit.2000.854108.

[75] N. Mohd. Noor, O. Mohd. Rijal, and C. Y. Fah, "Wavelet as features for Tuberculosis (MTB) using standard X-ray film images," *Int. Conf. Signal Process. Proceedings, ICSP*, vol. 2, no. August 2018, pp. 1138–1141, 2002, doi: 10.1109/ICOSP.2002.1179990.

[76] Y. Arzhaeva, L. Hogeweg, P. A. De Jong, M. A. Viergever, and B. Van Ginneken, "Global and local multi-valued dissimilarity-based classification: Application to computer-aided detection of tuberculosis," *Lect. Notes Comput. Sci. (including Subser. Lect. Notes Artif. Intell. Lect. Notes Bioinformatics)*, vol. 5762 LNCS, no. PART 2, pp. 724–731, 2009, doi: 10.1007/978-3-642-04271-3_88.

[77] R. Shen, S. Member, I. Cheng, S. Member, A. Basu, and S. Member, "A Hybrid Knowledge-Guided Detection Technique for Screening of Infectious Pulmonary Tuberculosis From Chest Radiographs," vol. 57, no. 11, pp. 2646–2656, 2010.

[78] J. Melendez *et al.*, "An automated tuberculosis screening strategy combining X-ray-based computer-aided detection and clinical information," *Sci. Rep.*, vol. 6, no. October 2015, pp. 1–8, 2016, doi: 10.1038/srep25265.

[79] S. Govindarajan, "Analysis of Tuberculosis in Chest Radiographs for Computerized Diagnosis using Bag of Keypoint Features," pp. 11–13, 2019.

[80] S. Hwang, H.-E. Kim, J. Jeong, and H.-J. Kim, "A novel approach for tuberculosis screening based on deep convolutional neural networks," in *Medical Imaging 2016: Computer-Aided Diagnosis*, 2016, vol. 9785, p. 97852W, doi: 10.1117/12.2216198.

[81] C. Liu *et al.*, "TX-CNN: Detecting tuberculosis in chest X-ray images using convolutional neural network," *Proc. - Int. Conf. Image Process. ICIP*, vol. 2017-Septe, pp. 2314–2318, 2018, doi: 10.1109/ICIP.2017.8296695.

[82] L. Devnath, S. Luo, P. Summons, and D. Wang, "Tuberculosis (Tb) Classification in Chest Radiographs using Deep Convolutional Neural networks.," vol. 6, no. June, pp. 68–74, 2018.

[83] S. J. Heo *et al.*, "Deep learning algorithms with demographic information help to detect tuberculosis in chest radiographs in annual workers' health examination data," *Int. J. Environ. Res. Public Health*, vol. 16, no. 2, 2019, doi: 10.3390/ijerph16020250.

[84] Y. Ban and H. Wang, "Rethinking Computer-aided Tuberculosis Diagnosis."

[85] U. Andayani, R. F. Rahmat, N. Sylviana Pasi, B. Siregar, M. F. Syahputra, and M. A. Muchtar, "Identification of the Tuberculosis (TB) Disease Based on XRay Images Using Probabilistic Neural Network (PNN)," *J. Phys. Conf. Ser.*, vol. 1235, no. 1, 2019, doi: 10.1088/1742-6596/1235/1/012056.

[86] *Global Tuberculosis Report*. 2019.

[87] M. Harris *et al.*, "A systematic review of the diagnostic accuracy of artificial intelligence-based computer programs to analyze chest x-rays for pulmonary tuberculosis," *PLoS One*, vol. 14, no. 9, p. e0221339, 2019, doi: https://doi.org/10.1371/journal.pone.0221339.

4

Forecasting Time Series Data Using ARIMA and Facebook Prophet Models

S. Sivaramakrishnan[1], Terrance Frederick Fernandez[2], R. G. Babukarthik[3], and S. Premalatha[4]

[1]Department of Electronics and Communication Engineering, Dayananda Sagar University, Bangalore, India
[2]Department of Information Technology, Dhanalakshmi Srinivasan College of Engineering & Technology, Chennai, India
[3]Department of Computer Science and Engineering, Dayananda Sagar University, Bangalore, India
[4]Department of Electronics and Communication Engineering, KSR Institute of Engineering and Technology, Tiruchengode, India
E-mail: sivaramkrish.s@gmail.com; frederick@pec.edu; r.g.babukarthik@gmail.com; shineprem1@yahoo.co.in

Abstract

Forecasting a parameter of interest in the time series data, based on the past values helps resource optimization in case of sensor networks, or to meet the demand of the society in case of analyzing essential commodity time series data or it helps in better analysis of a company value in near feature based on the revenue generated in the past. Auto Regression Integrated Moving Average (ARIMA) model is used to predict the future value of energy dissipation in case of sensor network, product, sales, share market fluctuation, essential commodity, etc. Facebook Prophet model is launched by Facebook which is an open-source library which can be used to analyze data which shows the variation in trend and seasonality. In this article, the number of passengers using aircraft is predicted using the ARIMA model and the Facebook Prophet model. Both the model performs well for the data considered and in ARIMA model, choosing the fundamental parameter for analysis of the data is challenging as compared to Facebook Prophet model as the latter uses in-built library function to ease the task.

Keywords: ARIMA, Forecasting, Facebook Prophet.

4.1 Introduction

ARIMA model is preferably used for analyzing the time series data and thereby forecast the future value. ARIMA model is a combination of Auto regression, Integrated, and Moving average models and hence abbreviated as ARIMA. Time series analysis of data is considering the parameter of interest in a data with respect to time like price of the commodity, weather condition over a year, variation of the stock price over certain period of time, etc. It also can be used in the analysis of the medical data like analyzing and predicting the heartbeat of a patient. Thereby, the application of time series analysis helps the industries to predict future growth or income of the organization, predicting the fluctuation of the stock price, analyzing the past data, and expecting its performance in the near future.

The time series analysis also helps in predicting the future demands of the commodity based on the trends pattern of the past, and thereby necessary planning can be done well in advance to meet the demand. This way, the essential needs of the society are always kept in track and necessary action can be taken to ensure that the requirement of the future can always be fulfilled. The time series analysis plays a major role in the block chain technology and is used by all major e-commerce technological giants like Amazon and Flip kart.

Facebook recently launched Prophet which is an open-source library which works well for the time series data. The key features of this model lie in handling the data with outliers and missing values. It can work efficiently with seasonality data involving the holidays and thus making it as a good choice for predicting the future value for time series analysis. The air passenger data from Kaggle is used here and the analysis is carried out with traditional ARIMA model and the same data is used for analysis using the Facebook Prophet model.

Some of the literatures are discussed as follows. Edgar S et al. used the probability and statistics over the data to predict future data [1]. Bendong Zhao et al. performed the time series analysis using Convolutional Neural Network (CNN) using the convolution and the pooling mathematical functions. The method also overcomes the effect of noise component [2].

Bin Yang et al. used a complex valued ordinary differential equation function for analyzing the time series data. The convergence speed of the proposed function is found to perform better than the standard techniques such as particle swarm optimization and crow search algorithms [3]. Zhengxin Li et al. designed a framework to check the similarity in the data which involves two stage analysis. The first stage involves analysis of the data to find out the series which is dissimilar and in the second stage more emphasis is carried out to reduce the computational cost [4].

Dio Li et al. used the bike sharing system data and it concentrates in reducing the dimension of the data by using a clustering technique to group the data set which have the similarity. The proposed algorithm also eliminates the effect of random noise [5]. Jin Fan et al. proposed an encoder–decoder framework where the data is gathered continuously with the help of sensor and the data is analyzed and the effect is made to understand the real time interaction [6]. Fagui et al. proposed an algorithm which mainly concentrates on two challenges faced by multivariate time series forecasting. The challenges are reducing the effect of the noise in the data and to predict the future data. The proposed algorithm has three stages and the first stage involves the decomposition of the data and the second stage involves analysis of the decomposed data and finally the last stage is ensemble [7].

Billy Tanuwijaya et al. proposed a neutrosophic hesitant fuzzy algorithm for predicting the future data in the time series analysis. The result shown the predicted value has less root mean square value [8]. Ya-nan wang et al. proposed an algorithm to check for any anomaly detection which find its application in cyber security. The algorithm uses the time series data to predict any anomaly detection in the network [9]. Chao Meng et al. concentrated in detecting the outlier in the time series data from real time system like sensor network, real time environmental data, etc. [10]. Zheng Zhang et al. proposed a time-adaptive optimal transport function which is an alternative to dynamic time wrapping and it works well on multiple datasets [11]. Federico Succetti et al. uses a different classification with deep neural network (DNN) to predict the photovoltaic energy and the application of the model helps in predicting the data for real time application and thus optimize the energy saving [12].

Burak Berk Ustundag et al. proposed a wavelet neural network for predicting the future data for the time series analysis. An additional network is also used to find out the main error and with this knowledge the overall error can be analyzed and corrected appropriately [13]. Tao Huamin et al. proposed a method to find out the missing values in the time series data using the appropriate technique. Since the missing values are not dropped, the analysis give more accurate results [14]. Fang-Mei Tseng et al. combines well known ARIMA and the neural network technique to predict the seasonal time series data. It is found that the error in predicting the data is reduced due to the combination of two models [15]. G. Peter Zhang proposed a hybrid model which involves both ARIMA and Artificial Neural Network (ANN) to predict the time series data. The advantage of this method is in predicting the time series data with better accuracy [16].

Gnacio Medina et al. present an application named Prophet. It is a web-based tool for prediction of time series data. The tool classifies the data into train data and test data and based on the training it will predict the result for any unseen data [17]. Toni Toharudin et al. proposed a hybrid long short-term memory and a Facebook Prophet model to predict the time series data for air temperature. The significance of Facebook Prophet model is that it can overcome the missing values in the data and can be used for analyzing the seasonal data and more importantly, it is open source [18]. Resa Septiani Pontoh et al. proposed a hybrid model consisting of Facebook Prophet and a feed forward neural network model. The main advantage of using the feed forward neural network is that it can handle huge real time data [19]. Mashael Khayyat et al. used Facebook Prophet and Python programming in predicting the Covid outbreak [20]. Miroslav Navratil et al. used the same Facebook Prophet model for analyzing the future forecast in the business [21].

4.2 Arima Model

The Moving Average (MA) model is used for predicting the value of the parameter for the time series data and it works good for the seasonal data. The MA models use the error value in the previous prediction and forecast the future value. The MA models can also be categorized as MA(1) and MA(2) based on the number of error values it takes to predict the new forecast. The advantage of this method is in predicting time series data with better accuracy in the previous forecast. Unlike the MA model in the Auto Regression (AR) Model, the forecast is done based on the past values of the parameter of interest. The linear combination of the past values is used to predict the future value.

The predicted value f' for MA(1) and MA(2) model is represented by the Equations (4.1) and (4.2) respectively

$$f' = m + \theta_1 \varepsilon_{t-1} \qquad (4.1)$$

$$f' = m + \theta_1\varepsilon_{t-1} + \theta_2\varepsilon_{t-2} \qquad (4.2)$$

Were $\varepsilon_{t-1}, \varepsilon_{t-2}$ represents the error in the previous forecast. Similarly, the MA n-th model is represented by the Equation (4.3)

$$y_t = c + \varepsilon_t + \theta_1\varepsilon_{t-1} + \theta_2\varepsilon_{t-2} + \cdots + \theta_n\varepsilon_{t-n} \qquad (4.3)$$

The AR model of order n can be represented as shown in the Equation (4.4)

$$y_t = c + \phi_1 y_{t-1} + \phi_2 y_{t-2} + \cdots + \phi_p y_{t-n} + \varepsilon_t \qquad (4.4)$$

Were $y_{t-1}, y_{t-2}, y_{t-n}$ represent the past predicted values and ε_t represents the overall error in the model. And the ARIMA model is represented by the Equation (4.5)

$$y_k = \sum_{i=1}^{k-1}(Z_{k-i} + y_n) \qquad (4.5)$$

were y_n is the last data which is available and Z_{k-i} represents the difference of data at two instances of time.

The model which integrates both AR model and MA model to predict the parameter of interest is called as the ARIMA model. The ARIMA model thus has three parameters: the first parameter is denoted by the letter 'p' indicating the order of Auto regression term, the second parameter is denoted by the letter 'q' which indicates the order of moving average term, and the final term is denoted by the letter 'd' and it is a number indicating the differencing required to make the time series stationary. It has to be noted that the data which has to be processing through the ARIMA model has to be checked for stationarity. If the data is not stationary, then necessary preprocessing has to be carried out to ensure the stationarity of the data. One such test to check whether the data is stationary or not is the Dicky–Fuller test.

4.2.1 Data Analysis Using ARIMA Model

The air traveler passenger list data is used to predict the future travelers using ARIMA model. The data consist of a total of 144 entries indicating the number of travelers for each month. The description of the data is shown in Figure 4.1. The number of passengers using the aircraft for various seasons is shown in Figure 4.2.

From the above Figure, it can be clearly visualized that, the mean is not stationary and the data has to be preprocessed to apply the ARIMA model.

	no_passengers
count	144.000000
mean	280.298611
std	119.966317
min	104.000000
25%	180.000000
50%	265.500000
75%	360.500000
max	622.000000

Figure 4.1 Data description.

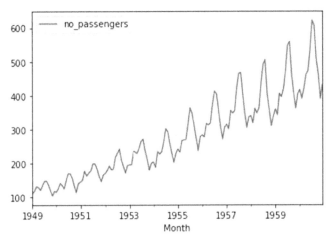

Figure 4.2 Number of passengers using aircraft.

The result for stationarity can also be got from Dicky–Fuller test as mentioned in Table 4.1 below.

Table 4.1 Result of Dicky–Fuller test without differencing

ADF test statistic	0.8153688792060543
p-value	0.9918802434376411
Lags used	13
Number of observations used	130

The p-value is the parameter used in Dicky–Fuller test and it should be lesser than 0.05 to be stationary and the results mentioned in the above table indicate that the data is not stationary. After applying the difference of 12 as it is a seasonal data and as it shows the variation for 12 months and then applying the test, the data is found to be stationary with p-value less than 0.05 as indicated by the Table 4.2 and Figure 4.3 which shows the visualization of the differenced data.

The ARIMA model is constructed with the values of p, q, d as 1, 1, 1 and since the data is seasonal, the seasonal order of 12 is chosen to predict the data from 130 to 143 entries and the result is as shown in the Figure 4.4.

Figure 4.4 shows that ARIMA model performs well as the forecast indicated by orange line is of close approximation to the existing data indicated by blue line. As the model holds good, the future data can be predicted and the result is shown in Figure 4.5 below

Table 4.2 Result of Dicky–Fuller test with differencing

ADF test statistic	−3.3830207264924805
p-value	0.0115514930855149
Lags used	1
Number of observations used	130

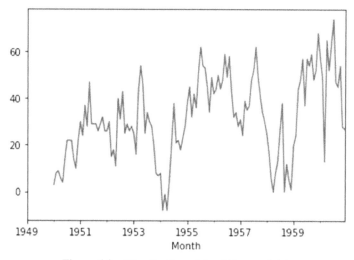

Figure 4.3 Visualization of the differenced data.

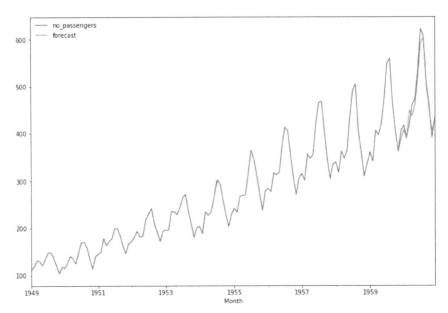

Figure 4.4 Data prediction using ARIMA (1,1,1) with seasonal order of 12.

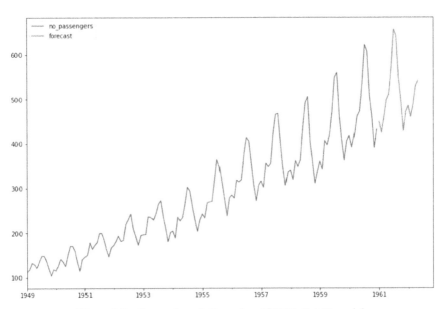

Figure 4.5 Forecast prediction using ARIMA (1,1,1) model.

4.3 Data Analysis Using Facebook Prophet Model

The Facebook Prophet model is used to forecast the data which involves the seasonal variation and these data hold good for analysis using this model. The Facebook Prophet is an open-source library which works for analyzing the time series data. The same data used in ARIMA model is used in Prophet model and the data has to be preprocessed before applying to the Prophet model. The parameters which are not required are removed from the data and then date-time format is applied to the column holding the date details which is necessary to carry out the analysis in the Facebook Prophet model. The advantage of this model is that when the data is fitted to the model, it could automatically detect the type of seasonality. In the data considered, the model accurately predicts that the seasonality holds good for yearly analysis.

The prediction is carried out for the next one year and the model learns from the existing data and does the prediction for the future. Figures 4.6 and 4.7 below show the predicted analysis for various dates.

The value mentioned under the column yhat is the predicted values for the dates mentioned in Figure 4.6 under the column ds. Figure 4.8 below shows the plot indicating the prediction for the future dates, where the black dots indicate the actual value and the blue lines indicate the predicted values.

	ds	trend	yhat_lower	yhat_upper	trend_lower	trend_upper	additive_terms	additive_terms_lower	additive_terms_upper	yearly	yearly_lower
0	1949-01-01	106.320611	55.777817	114.880449	106.320611	106.320611	-21.941934	-21.941934	-21.941934	-21.941934	-21.941934
1	1949-02-01	108.500995	49.551448	108.894815	108.500995	108.500995	-30.714169	-30.714169	-30.714169	-30.714169	-30.714169
2	1949-03-01	110.470374	80.974581	139.573971	110.470374	110.470374	-0.475833	-0.475833	-0.475833	-0.475833	-0.475833
3	1949-04-01	112.650758	77.592021	138.202184	112.650758	112.650758	-5.203855	-5.203855	-5.203855	-5.203855	-5.203855
4	1949-05-01	114.760808	82.127508	139.129781	114.760808	114.760808	-3.825854	-3.825854	-3.825854	-3.825854	-3.825854

Figure 4.6 Prediction analysis for future dates.

yearly_lower	yearly_upper	multiplicative_terms	multiplicative_terms_lower	multiplicative_terms_upper	yhat
-21.941934	-21.941934	0.0	0.0	0.0	84.378677
-30.714169	-30.714169	0.0	0.0	0.0	77.786826
-0.475833	-0.475833	0.0	0.0	0.0	109.994541
-5.203855	-5.203855	0.0	0.0	0.0	107.446903
-3.825854	-3.825854	0.0	0.0	0.0	110.934953

Figure 4.7 Prediction analysis for future dates.

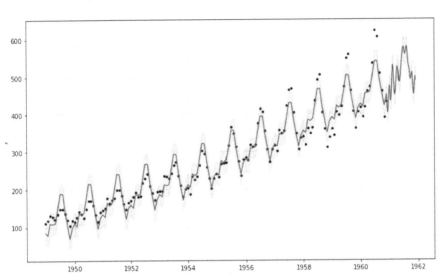

Figure 4.8 Prediction using Facebook Prophet Model.

The trend prediction and the yearly prediction are shown in Figure 4.9 below.

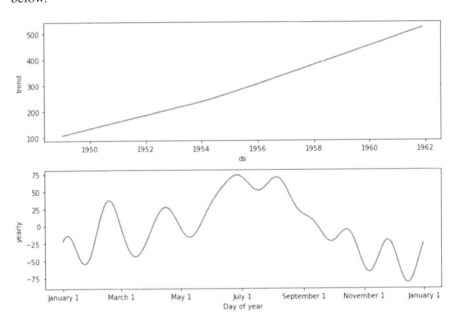

Figure 4.9 Trend and yearly prediction.

4.4 Conculsion

Both the ARIMA model and the Facebook Prophet model work well for the same data which is considered in this article. Choosing the correct ARIMA model involves selecting the values of p, q, and d. Wrong choice of these parameters might not result in the accurate prediction and the main challenge of ARIMA model is to find the right values of these parameters. On the other hand, the Facebook Prophet model can analyze the data using the in-built function in the library and clearly predict the type of seasonal data such as weekly seasonal, monthly seasonal, or yearly seasonal. In the data considered, it is a yearly seasonal and based on that seasonal prediction the model does the necessary analysis and predicts the data more accurately.

References

[1] García-Treviño, Edgar S., and Javier A. Barria. "Structural generative descriptions for time series classification." IEEE transactions on cybernetics 44, no. 10 (2014): 1978–1991.

[2] Zhao, Bendong, Huanzhang Lu, Shangfeng Chen, Junliang Liu, and Dongya Wu. "Convolutional neural networks for time series classification." Journal of Systems Engineering and Electronics 28, no. 1 (2017): 162–169.

[3] Yang, Bin, and Wenzheng Bao. "Complex-valued ordinary differential equation modeling for time series identification." IEEE Access 7 (2019): 41033–41042.

[4] Li, Zhengxin, Jiansheng Guo, Hailin Li, Tao Wu, Sheng Mao, and Feiping Nie. "Speed up similarity search of time series under dynamic time warping." IEEE Access 7 (2019): 163644–163653.

[5] Li, Duo, Yifei Zhao, and Yan Li. "Time-series representation and clustering approaches for sharing bike usage mining." IEEE Access 7 (2019): 177856–177863.

[6] Fan, Jin, Hongkun Wang, Yipan Huang, Ke Zhang, and Bei Zhao. "AEDmts: An Attention-Based Encoder-Decoder Framework for Multi-Sensory Time Series Analytic." IEEE Access 8 (2020): 37406–37415.

[7] Liu, Fagui, Yunsheng Lu, and Muqing Cai. "A Hybrid Method With Adaptive Sub-Series Clustering and Attention-Based Stacked Residual LSTMs for Multivariate Time Series Forecasting." IEEE Access 8 (2020): 62423–62438.

[8] Tanuwijaya, Billy, Ganeshsree Selvachandran, Mohamed Abdel-Basset, Hiep Xuan Huynh, Van-Huy Pham, and Mahmoud Ismail. "A Novel Single Valued Neutrosophic Hesitant Fuzzy Time Series Model: Applications in Indonesian and Argentinian Stock Index Forecasting." IEEE Access 8 (2020): 60126–60141.

[9] Wang, Ya-Nan, Jian Wang, Xiaoshi Fan, and Yafei Song. "Network Traffic Anomaly Detection Algorithm Based on Intuitionistic Fuzzy Time Series Graph Mining." IEEE Access 8 (2020): 63381–63389.

[10] Meng, Chao, Xue Song Jiang, Xiu Mei Wei, and Tao Wei. "A Time Convolutional Network Based Outlier Detection for Multidimensional Time Series in Cyber-Physical-Social Systems." IEEE Access 8 (2020): 74933–74942.

[11] Zhang, Zheng, Ping Tang, and Thomas Corpetti. "Time Adaptive Optimal Transport: A Framework of Time Series Similarity Measure." IEEE Access 8 (2020): 149764–149774.

[12] Succetti, Federico, Antonello Rosato, Rodolfo Araneo, and Massimo Panella. "Deep Neural Networks for Multivariate Prediction of Photovoltaic Power Time Series." IEEE Access 8 (2020): 211490–211505.

[13] Ustundag, Burak Berk, and Ajla Kulaglic. "High-Performance Time Series Prediction With Predictive Error Compensated Wavelet Neural Networks." IEEE Access 8 (2020): 210532–21054.

[14] Huamin, T. A. O., D. E. N. G. Qiuqun, and X. I. A. O. Shanzhu. "Reconstruction of time series with missing value using 2D representation-based denoising autoencoder." Journal of Systems Engineering and Electronics 31, no. 6 (2020): 1087–1096.

[15] Tseng, Fang-Mei, Hsiao-Cheng Yu, and Gwo-Hsiung Tzeng. "Combining neural network model with seasonal time series ARIMA model." Technological forecasting and social change 69, no. 1 (2002): 71–87.

[16] Zhang, G. Peter. "Time series forecasting using a hybrid ARIMA and neural network model." Neurocomputing 50 (2003): 159–175.

[17] Medina, Ignacio, David Montaner, Joaquín Tárraga, and Joaquín Dopazo. "Prophet, a web-based tool for class prediction using microarray data." Bioinformatics 23, no. 3 (2007): 390–39.

[18] Toharudin, Toni, Resa Septiani Pontoh, Rezzy Eko Caraka, Solichatus Zahroh, Youngjo Lee, and Rung Ching Chen. "Employing long short-term memory and Facebook prophet model in air temperature forecasting." Communications in Statistics-Simulation and Computation (2020): 1–24.

[19] Pontoh, Resa Septiani, S. Zahroh, H. R. Nurahman, R. I. Aprillion, A. Ramdani, and D. I. Akmal. "Applied of feed-forward neural network and facebook prophet model for train passengers forecasting." In Journal of Physics: Conference Series, vol. 1776, no. 1, p. 012057. IOP Publishing, 2021.

[20] Khayyat, Mashael, Kaouther Laabidi, Nada Almalki, and Maysoon Al-zahrani. "Time Series Facebook Prophet Model and Python for COVID-19 Outbreak Prediction." CMC-COMPUTERS MATERIALS & CONTINUA 67, no. 3 (2021): 3781–3793.

[21] Navratil, Miroslav, and Andrea Kolkova. "Decomposition and Forecasting Time Series in the Business Economy Using Prophet Forecasting Model." Central European Business Review 8, no. 4 (2019): 26.

5

A Novel Technique for User Decision Prediction and Assistance Using Machine Learning and NLP: A Model to Transform the E-commerce System

V. Vivek[1,*], T. R. Mahesh[1], C. Saravanan[1], and K. Vinay Kumar[2]

[1]Faculty of Engineering and Technology, Department of Computer Science and Engineering, JAIN (Deemed to be University), Bangalore, India
[2]Department of Computer Science and Engineering, KITS Warangal, Telangana, India
E-mail: v.vullikanti@jainuniversity.ac.in; t.mahesh@jainuniversity.ac.in; c.saravanan@jainuniversity.ac.in; kvk@cse.kits.ac.in
*Corresponding Author

Abstract

Rapid developments in the usage of Internet enabled the need of predictive analytics and assistance to understand the behavior of buyers and sellers on e-commerce platform. Online shopping has evolved in many years since its inception, and also inherited lot of advancements in the way we live, shop, and do business. Finding potential new buyers, retargeting the existing one, handling queries, and maintaining the inventories are very challenging modules and will affect the e-commerce industry very badly if not taken care of. Artificial Intelligence (AI) is being used by top e-commerce industries like Amazon and Flipkart to understand the behavior of the buyer by introducing chatbots and customized recommendations by predicting the shopping pattern of the buyers. These top industries maintain and process millions of buyer behaviors to predict and target the right customers. A novel technique called Call based Intelligent Bot Personal Assistance (CIB-PA) using Machine Learning (ML) and Natural Language Processing (NLP) has been proposed in the paper to self-serve the seller in assisting and targeting the right buyer more accurately. CIB-PA is deployed by integrating text analysis via call recording, Intelligence as a Service (IasS), Personal Assistance, and Bot Integrations. Test bed was deployed using App development toolkits, Hadoop MapReduce, DataRobot, RapidMiner, Fusioo, BigML, and CoreNLP. Through simulations, it is observed that CIB-PA outperforms in predicting buyer behavior.

Keywords: Artificial Intelligence, Machine Learning, Natural Language Processing, Predictions, Recommended System, E-commerce.

5.1 Introduction

Cross-selling products on e-commerce platforms like Amazon, Flipkart, and Snapdeal are the major sources of profit for the sellers. There are various technologies and approaches that boosted these platforms to attract the retailers and buyers. Prediction and recommendation systems using Artificial Intelligence (AI) are the models that are primarily opted by the retailers to gain more profit. Beside this, AI is also been opted to provide chatbot services, attracting potential new and existing customers in buying related products. Various studies have projected that online sales will be reaching $4.8 billion this year. Gartner, one of the leading prediction companies, has predicted that nearly 80% of the buyer's interactions will be based on AI frameworks. AI has transformed the buyer experience through the prediction and recommendation systems. For an example, if a buyer buys a specific brand of milk on a regular basis, then the seller may send him a personalized offer or he may suggest the buyer few more items which goes well with the milk. Out of multiple applications of AI for e-commerce, top four applications are chatbot service, customized personal recommendation, optimizing the warehouse activities, and image-based product discovery. Deep learning using NLP is another combination that makes dream come true for the sellers to focus on individual buyers [1, 2].

In 1969, the first ever e-commerce service, CompuServe was introduced by electrical engineering students using dial-up connection. In 1979, electronic shopping using customized TV to perform transactions with the help of telephone line was deployed by an English inventor named Michael Aldrich. The first e-commerce industry that entered the market was Boston Computer Exchange in 1982. Books were sold online in 1992 by Book Stacks using dial-up bulletin format and later the company has switched to go live on Internet using their own domain. Amazon entered the e-commerce market to sell books in 1995. Focusing on the rapid changes in e-commerce platform, PayPal introduced money transfer tool in 1998 and later joined hands with Elon Musk's online banking organization. With $25 million funding, the giant industry Alibaba has entered online marketplace in 1999 and earned profits. Looking at the changes toward online shopping and demands, in 2000, Google introduced the concepts of AdWords which helped many e-commerce portals to advertise their products via Google search. The biggest advancement to sell a variety of things online was released in the year 2004 with Shopify, in the year 2005 by Amazon Prime membership and Etsy.

By linking with the buyer's bank account, digital payments using e-wallets option was introduced by Google in 2011, and Apple Pay was introduced in 2014 to perform mobile payments. Focusing on profits and peer competitions, e-commerce has rapidly inherited various algorithms and technologies deployed by AI, ML, NLP, Deep learning, User Interface (UI), and User Experience (UX) [1–4].

Small- to large-scale retailers have focused on e-commerce platforms and expanded their networks to reach many customers. Platform has very big support for the small businesses to reach their buyers directly and opening a new door for profits. The impact of e-commerce is remarkable on small businesses, where the owners have stated to host their own platforms with various offers and payment methods. The rise of e-commerce market place is shown in Figure 5.1.

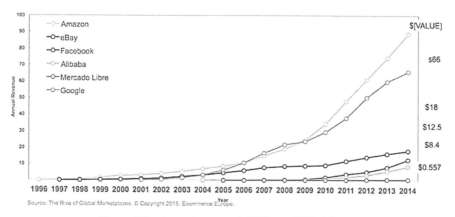

Figure 5.1 Annual revenue in billions of US dollars.

Introducing variety of products from different brands has scaled up the revenues for the retailers. AI and very significantly ML which is a subset of AI have an extreme impact on business. ML can enable smart search using NLP, using previous search history, the algorithm can suggest what users genuinely wants when one starts online shopping [5]. Recommendations provided by the ML algorithms are much smarter; they can analyze the behavior of the buyer so accurately, where it can suggest the items based on returns made by the users.

The aim of this paper is to deploy a novel technique called CIB-PA which is designed using ML and NLP. The CIB-PA predicts the buyer interests and needs using call recordings. This will be a great patch to the existing e-commerce retailers to reach buyers more accurately and intelligently. The rest of the sections are focused as follows: Section 5.2 gives a detailed summary of various existing AI and its related area approaches and models for e-commerce platform; Section 5.3 describes the proposed CIB-PA model; Section 5.4 examines the proposed model behavior compared to other existing models; and Section 5.5 concludes the behavior of the CIB-PA with further challenges and improvements.

5.2 Related Work

Machine learning enables the retailers to design more customized buyer experiences. There are proven case studies where ML has reduced the buyer issues before they have encountered any issues. This has drastically reduced the personal chart rejection rates and increased the sales. Chatbots were able to provide unbiased service and solutions for the buyer issues. Shanshan Y et al. proposed a recommended system for analyzing the buyer sentiment using the feedback and review [6]. The recommended system proposed by the author focused on collaborative filtering, content-based filtering, similarity based, and matrix factorization. Figures 5.2 and 5.3 give system model and proposed model used for testbed simulation. The proposed model was simulated to self-understand the customer online shopping data such as reviews, patterns, and propose product for the buyer with 98% accuracy.

Another method proposed by authors Giorgos and Rigas using UC Irvine's Machine Learning Repository data has been analyzed to understand the intention of the buyer. The collection of data covers the browsing history, location, type of packets, existing or new buyer, and time of the visit. Zeinab Shahbazi et al. [7] proposed XGBoost-based item recommendation system using various small data sets. Authors collected raw data from various clicks

5.2 Related Work 65

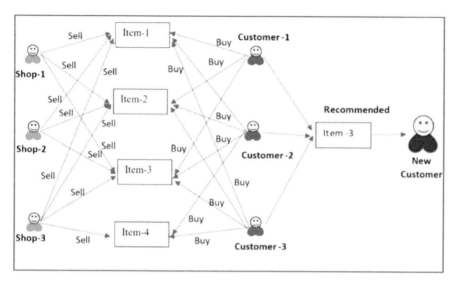

Figure 5.2 Shanshan Y et al., collaborative recommendation based on product recommendation.

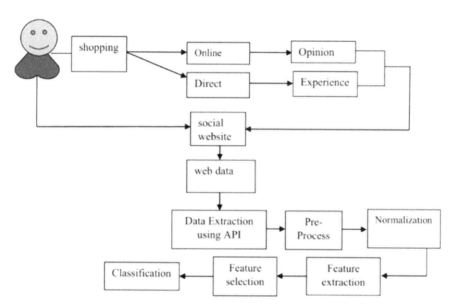

Figure 5.3 Shanshan Y et al., proposed model.

and supplied as input to XGBoost algorithm deployed using machine learning. XGBoost summarizes the overall prediction rate and recommends top five items to the buyer.

Ravali et al. discussed about the importance of text summarization and various methods available for text summarization so that buyer can have short and genuine review about the product that he or she is interested to buy [8]. Authors classified the text summarization in to three models, firstly based on the input type, i.e., single or multiple documents; secondly based on the purpose, i.e., generic or domain specific or query-based; and finally based on the output type, i.e., extractive or abstractive. Authors have recommended Seq2Seq method along with long short-term memory for accuracy in prediction.

Yang et al. in their work focused on designing methodology (q−ROFIWHM) for online buyers with the help of deep learning [9]. Using deep learning model at first level q−ROFIWHM extracts the product attributes and their reviews to match the respective opinion pairs along with the sentiment mapping. q−ROFIWHM calculates proportions using three different types of sentiments; there by this sentiment information is transformed to multiple cross-decision tables with the help of q-rung fuzzy set. Finally, using these multiple cross-decision tables, q−ROFIWHM summarizes the cross-decision information and gives ranking to the result for supporting buyer decision. Hemalatha et al. proposed a framework to determine the implicit analysis of reviews along with Apriori algorithm [10]. On the input data set, natural language processing tokenization is applied as an initial step. Data is moved further through Apriori algorithm, sentimental analysis, deep learning methods, deep parsing, explicit rule, and finally implicit rule.

Wei et al. designed a framework to predict the item performance using text emotions with the help of NLP and VAR [11]. Authors work is focused to predict the retailer performance and buyer behaviors for specific products using Amazon data set. Bidirectional Encoder Representations from Transformers (BERT) is designed using NLP, to understand and predict the test rating with help of pair behavior patterns. Authors recoded a 20% accuracy improvement compared with traditional BERT. PCA model has been integrated with extracted reviewer's popularity, item reputation, and other attributes. Finally, pacifier vector autoregression (VAR) is integrated and determined the impulse response and variance.

Zeng et al. simulated e-commerce activities taken place during a large shopping festival in China [12]. To improve retailers' profits, buyer's satisfaction, and effective warehouse management for product delivery, authors

5.2 Related Work

have analyzed 31 million logs during peak sales time. A collaborative filtering technique is applied to predict, recommend, and see whether buyers have purchased the product or not. Authors observed that as shown in Figure 5.4 and in Table 5.1 only 9.80% of users searched for a product and thereby added to cart and proceeded for purchase; 75.2% buyers only searched and reviewed; and 11.4% buyers searched for an item and added it to cart but never bought it.

The summary of existing methods and techniques has focused on buyer behavior at online platforms, retailer performance, and buyer reviews. It is also noted that the existing models do not support simulating implicit and explicit unique products in a customized environment, as a result there is a need for a model which can predict more intelligently and accurately though integrating multi-layer attributes. The proposed work CIB-PA focuses on understanding the buyer requirement in a unique way of call analyzing and gives more accurate and proactive recommendations to the customers.

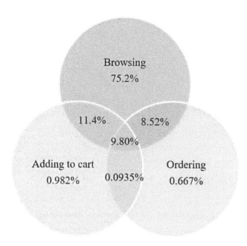

Figure 5.4 Observations of Zeng et al. on buyers' online shopping behavior.

Table 5.1 Summary of buyer's actions on online products by Zeng et al.

Action Type	Users	Products
Browsing	47,124 (98.4%)	226.355 (95.6%)
Adding to cart	10,011 (20.9%)	17,512 (7.39%)
Ordering	8568 (17.9%)	10,808 (4.56%)

5.3 Research Methodology

The proposed work aims to advance the approach of predicting buyer requirements by recording the buyer call conversations. The work integrates the deep parsing and NLP to record the buyer communications and analyze the same. As an example scenario, consider a buyer b_i is in need of purchasing a laptop for his daily usage and he rings up his close friend to ask for suggestions and recommendations. The pre-installed application on buyer mobile records the whole communication up on approved permission from the buyer (we may refer buyer as user whenever it is required). This application can be disabled by the user whenever he or she needs based on individual privacy settings. Upon recording the user conversation (could be of 10 minutes or 1 hour duration), the complete recording is stored in either cloud or local storage. The stored recording is then processed and converted into text for cleaning purpose.

The cleaned text summarizes the unique keywords and predicts the user requirement. Immediately after this stage, the CIB-PA algorithm performs mapping with different e-commerce products and gives a recommendation to the user about product availability, price, and expected date of availability, if not available and other recommendations as notification to the user. This model can server a plug-in to the existing e-commerce retailers, assume Flipkart integrate this plug-in and recommend the product to it users as notification. Users have not performed any implicit search on the platform, but still the algorithm attempts to predict the user requirement and recommend the products. Why not! It can be an application with integrated CIB-PA and can process user call conversation (to meet some "abc" person on a specific date and time) and suggest me local offers available and request me to pre-book the table and items. There are multiple stages the call recording is processed through as given in Figure 5.5.

Initially, the call recording was limited to audio sources like headset microphones. Recent advancements in the architecture of mobile devices enable three new features called voice uplink, downlink, and call, by which one can attempt to record the call. Through customization at different SDL levels, call recording is possible via uplink and downlink continuously. By integrating GRADLE, voice call is used for audio source for recording the selected, incoming and outgoing calls were automatically recorded by mapping contact name or number to the recording file. Sample Java patch logic is given below.

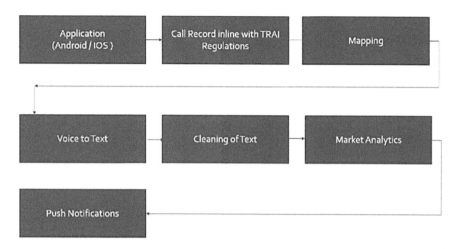

Figure 5.5 CIB-PA process flow for user requirement prediction and recommendation.

Patch to Request Recording Permission

public void onRequestPermissionsResult(int requestCode, String permissions[], int[] grantResults)

{

  switch (requestCode)

    {

      case REQUEST_STORAGE_PERMISSION:

if (grantResults.length > 0 &&

PackageManager.PERMISSION_GRANTED == grantResults[0])

{

Utility.setRecordingSkipMediaScan(mPrefSkipMediaScan.isChecked());

}

break;

  default:

  break;

        }

    }

Patch to Call recording

findAndHookMethod(callRecordingServiceName, lpvivek.classLoader, "isEnabled", Context.class1, new XC_MethodHook1()

{

  @Override_1

    protected void afterHooked1Method   (XC_MethodHook1.MethodHook1vivek vivek) throws Throwable

{

  sSettings.reload();

  if (sSettings.isRecordEnable())

  {

      vivek.setResult(Boolean.TRUE);

  }

}

});

Mozilla DeepSpeech recognition is used to convert the speech to text, the default WitAi's free interface is integrated for audio reorganization and conversation. At times, for testbed simulations, Google's Speech-to-Text API is integrated for accuracy companions. A built-in package called languageCode is used in cases where user speaks languages other than English. Whenever Google's Speech-to-Text API is used, a new service account key will be generated to integrate JSON patch file to the model. Figure 5.6 demonstrates the process of speech to text flow.

The content (reviews and comments) generated by e-commerce portals are often inappropriate. Even the user call recording contains repeated and out of the box discussions. It is very essential to clean such content and identify the key objectives to create a normalized text representation according to user requirements. Ftfy and unidecode instances are used to process the raw data. Python unicode package for data normalize also integrated for transliteration, due to the fact that manual mapping is enabled though it is superior. Customized functionalities and scripts will scan raw text attributes and report the potential key words. These potential key words are analyzed to reframe as normalized sentence with more accurate meaning. Row filtering, subbing, replacement, incomplete sentence, slang, word elongation, tokens, time stamps, symbols, ordinal numbers, and other few attributes are key factors while cleaning text. Using lexicon lookup

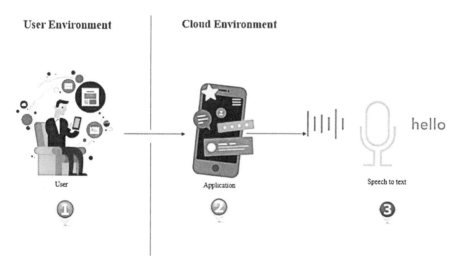

Figure 5.6 Model for speech to text using mobile application.

Out-Of-Vocabulary (OOV) detection is carried out, and also based on similarity function user enumeration was progressed by relating the noisy tokens of the lexicon.

Based on the key attributes collected from the cleaning phase, market analysis is carried out to list the feasible recommendations to the user. EC2 instance is used to collect the data sets from different sources, by invoking local container this instance data is loaded to containers for further processing. With the help of Hadoop Spark and these containers CIB-PA functions as recommendation model. There by the CIB-PA push the notifications to the user as recommendations and suggestions.

5.4 Experimental Results

Our data sets contain user call recordings from authorized users who have installed the customized application for processing the recording. Initial testbed was focused on recording discussions made for laptop purchase, and later extended to other attributes. In order to provide the accurate recommendations, first, we focused on recordings with 30-min duration having more than 3000 words. The text cleaning was done effectively to optimize the discussion and record the summary. Next, we excluded the out-of-discussion points to normalize the sentence to the key point. Survey was carried out to understand the accuracy of the recommendation. Table 5.2 presents the survey results.

Figure 5.7 gives the comparison of CIB-PA model and existing models, where the orders have been increased by 6% based on the recommendations pushed by the application.

Comparison of number of recommendations based on precision and recall is given in Figure 5.8. Improvement in the number of recommendations will always have an increase impact, at the same time decrease in the precision due to lower predicted ratings are considered in a recommendation.

Table 5.2 The survey results of the buyer recommendation accuracy

Recommendation Method	Average	Standard Deviation
Randomly-chosen model	3.76	1.342
Proposed model (CIB-PA)	4.51	1.010

5.4 Experimental Results 73

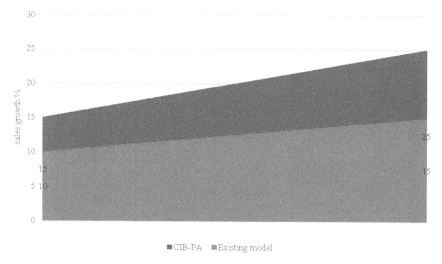

Figure 5.7 Comparison of CIB-PA model and existing sales.

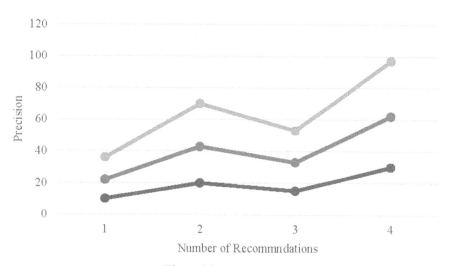

Figure 5.8 (a) Precision.

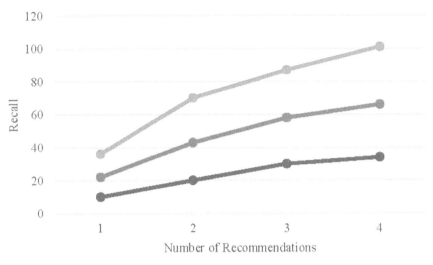

Figure 5.8 (b) Recall.

5.5 Conclusion and Future Scope

CIB-PA model has been deployed successfully to record the user call with recommended permissions and approval. Recording has been converted to text more accurately and thereafter the cleaning is done to remove inappropriate text, out of the topic conversations, and incomplete communications. Key attributes were listed more accurately and web crawling was done successfully to predict and recommend the suggestion to the buyer. But at the same time, it is observed that the retailers should maintain a patch to reply the server call to give more accurate and real-time data which will support predictions. Since it is not so easy to get real-time data from the various retailers on lively basis, this has been taken care using a hybrid patch model for filtering and providing the real-time data. We have conducted multiple experiments, which have led to the below conclusion. First, call recording with various time durations has been recorded and stored in the cloud environment, especially when size of the recording is very high in volume. Secondly, CIB-PA is able to clean the converted text with 94% accuracy, which is an 80% increase compared to the existing models. Further, a multi-layer approach can be adopted to save the processing time; this can be possible with storing the recording partially in local device, cloud, and fog environment.

References

[1] C. O. Sakar, S. O. Polat, M. Katircioglu, and Y. Kastro, "Real-time prediction of online shoppers' purchasing intention using multilayer perceptron and LSTM recurrent neural networks," *Neural Comput. Appl.*, vol. 31, no. 10, pp. 6893–6908, 2019, doi: 10.1007/s00521-018-3523-0.

[2] F. Shirazi and M. Mohammadi, "A big data analytics model for customer churn prediction in the retiree segment," *Int. J. Inf. Manage.*, vol. 48, no. February, pp. 238–253, 2019, doi: 10.1016/j.ijinfomgt.2018.10.005.

[3] P. Golnar-Nik, S. Farashi, and M. S. Safari, "The application of EEG power for the prediction and interpretation of consumer decision-making: A neuromarketing study," *Physiol. Behav.*, vol. 207, no. April, pp. 90–98, 2019, doi: 10.1016/j.physbeh.2019.04.025.

[4] W. Fang, Y. Guo, W. Liao, K. Ramani, and S. Huang, "Big data driven jobs remaining time prediction in discrete manufacturing system: a deep learning-based approach," *Int. J. Prod. Res.*, vol. 58, no. 9, pp. 2751–2766, 2020, doi: 10.1080/00207543.2019.1602744.

[5] M. Trupthi, S. Pabboju, and N. Gugulotu, *Deep sentiments extraction for consumer products using NLP-based technique*, vol. 898. Springer Singapore, 2019.

[6] S. Yi and X. Liu, "Machine learning based customer sentiment analysis for recommending shoppers, shops based on customers' review," *Complex Intell. Syst.*, vol. 6, no. 3, pp. 621–634, 2020, doi: 10.1007/s40747-020-00155-2.

[7] Z. Shahbazi and Y. Byun, "Product Recommendation Based on Content-based Filtering Using XGBoost Classifier Product Recommendation Based on Content-based Filtering Using XGBoost Classifier," *Int. J. Adv. Sci. Technol.*, vol. 29, no. July, pp. 6979–6988, 2020.

[8] R. Boorugu and G. Ramesh, "A Survey on NLP based Text Summarization for Summarizing Product Reviews," *Proc. 2nd Int. Conf. Inven. Res. Comput. Appl. ICIRCA 2020*, pp. 352–356, 2020, doi: 10.1109/ICIRCA48905.2020.9183355.

[9] Z. Yang, T. Ouyang, X. Fu, and X. Peng, "A decision-making algorithm for online shopping using deep-learning–based opinion pairs mining and q-rung orthopair fuzzy interaction Heronian mean operators," *Int. J. Intell. Syst.*, vol. 35, no. 5, pp. 783–825, 2020, doi: 10.1002/int.22225.

[10] B. Hemalatha and T. Velmurugan, "Direct-Indirect Association Rule Mining for Online Shopping Customer Data using Natural Language Processing," *Int. J. Recent Technol. Eng.*, vol. 8, no. 4, pp. 11099–11106, 2019, doi: 10.35940/ijrte.d7396.118419.

[11] Z. Wei, B. Han, and E. Zhou, "Research on text emotion analysis and product performance based on NLP and VAR model," *IOP Conf. Ser. Earth Environ. Sci.*, vol. 632, no. 5, 2021, doi: 10.1088/1755-1315/632/5/052077.

[12] M. Zeng, H. Cao, M. Chen, and Y. Li, "User behaviour modeling, recommendations, and purchase prediction during shopping festivals," *Electron. Mark.*, vol. 29, no. 2, pp. 263–274, 2019, doi: 10.1007/s12525-018-0311-8.

6
Machine Learning-Based Intelligent Video Analytics Design Using Depth Intra Coding

Kumbala Pradeep Reddy[1], Sarangam Kodati[2], Thotakura Veeranna[3], and G. Ravi[4]

[1]Department of CSE, CMR Institute of Technology (Autonomous), Hyderabad, Telangana, India
[2]Department of CSE, Teegala Krishna Reddy Engineering College, Telangana, India
[3]Department of CSE, Sai Spurthi Institute of Technology, Sathupally, Telangana, India
[4]Department of CSE, MRCET, Hydarabad, Telangana, India
E-mail: pradeep529@gmail.com; k.sarangam@gmail.com; veeru38@gmail.com; g.raviraja@gmail.com

Abstract

Machine learning-based intelligent video analytics design using depth intra coding is implemented. There is a requirement for an intelligent and smart system that can solve this deficiency by automating the process. When unsecure objects in the frame are detected, then this should give an alert to the security person. This approach restores depth component for conventional intra-picture coding. Preanalysis of coding unit is done for giving input. After preanalysis, the features are selected from the obtained data. The fast depth intra coding mode and 2N × 2N intra mode will perform the operation. If the depth or size is not equal, then it will again perform its operation from the preanalysis. If the partition is equal, then it will depend on the increment of depth and if it is not equal then content decision will process the screen content. If screen content is not equal, then it will perform the N × N intra mode and if it is equal then it performs Depth Intra Coding and PLT mode. Finally, best Coding Unit depth intra mode is chosen. If the depth, size is not equal then it will again perform its operation from the pre analysis. Hence, from the results it can observe that it gives effective results.

Keywords: Depth Intra Coding (DIC), Coding Unit (CU), Depth Size, Machine Learning, Feature Collection, Bjonteggard Delta Bit Rate (BDBR), Signal to Noise Ratio, Complexity, Accuracy.

6.1 Introduction

Picture or video coding is a rule which suggests to the processing innovation that packs picture or video into double code (e.g., bits) to work with capacity and transmission. The pressure could possibly guarantee ideal reproduction of picture/video from the pieces, which is named lossless and lossy coding individually. For common picture or video, the pressure proficiency of lossless coding is typically underneath necessity, so the majority of endeavors are dedicated to lossy coding [1]. Lossy picture or video coding arrangements are assessed at two perspectives: first is the pressure proficiency, usually estimated by the quantity of pieces (coding rate), minimizing would be ideal; second is brought about errors, generally estimated by the nature of the remade picture/video contrasted and the first picture or video. Picture or video coding is a basic and empowering innovation for PC picture preparing, PC vision, and visual correspondence.

The creative work of picture/video coding can be followed right back which exactly on schedule as present day imaging, picture dealing with, and visual correspondence systems. For example, Picture Coding, a grand overall get-together offered expressly to types of progress in picture or video coding [2, 3]. In this, different undertakings from both insightful world and industry have been given to this field.

Metropolitan zones or current urban areas are for the most part presented to tremendous group and traffic. Such zones needed to be under consistent re-consistence for security purposes. Past techniques like manual perception by security individual 24 × 7 is anything but a successful or practical strategy for re-consistence. However, if any individual conveys a blade or an arm in a jam-packed spot, it may not be identified because of the lack of ability of the security individual to discover such items in the group by the unaided eye. Subsequently, we require a long-lasting framework which will tackle this issue, a framework which is unequipped for including the quantity of individuals in the casing and separate among safe and unstable articles. 3D Video innovation has appeared as of late alongside expanded exploration at all phases of the preparing chain from 3D video catch to the presentation innovation. This will gradually incorporate new and progressed 3D video coding techniques for powerful pressure and transmission and furthermore most recent applications that blend 3D PC designs components and 3D recordings.

Here, the Multi View Video (MVV) portrayal design is used, giving similar scenes from at least two separate points of view. We notice tremendous group and traffic in metropolitan territories. These zones need constant reconnaissance for security reasons. A standard technique that utilizes manual perception by a security individual 24×7 is certainly not a successful or reasonable path for reconnaissance. Envision if any individual conveys a blade or weapon in a stuffed spot, it may not be recognized in light of the insufficiency of the security faculty to notice such items in the group by the unaided eye [4].

Along these lines, there is essential of a system which would incorporate the amount of individuals and separated among protected articles. Hence, this might diminish the load on the security work power and cooperation the perception more useful and suitable. The possibility of uniqueness is utilized to see conditions which are effectively used with MVC. This provides tremendous coding when diverged from simulcast coding. This idea is further appropriate to the HEVC standard for introducing a straightforward sound system and multi view video coding expansion [5]. The better coding effectiveness of HEVC contrasted with H.264/AVC is acquired. Compelling pressure and transmission of MVD information is the significant angle for the accomplishment of the model. The significance of intra coding strategy with another bearing and expecting mixing of 3D video and 3D PC plans [6]. The consistent and convincing coding plan is the key achievement for 3D video applications that either used assertion guides or polygon organizations, with further developed design blocks for a high assertion coding practicality in thought of least trinkets in outfitted points of view with an immaterial number of triangles of cross area extraction for a referred to bits rate [7, 8].

The new creations are the lattice extraction calculation that coordinates effectively in decoder for uncommonly planned arrangement of displaying capacities and applicable expecting modes for the improvement of the unaltered idea for totally dependent coding of the statement related intra encoder and decoder at all the stages. The primary point is the sign of a profundity block approach by mathematical sources planning capacities that permit introduction of scene surface with a least number of triangles.

6.1.1 Object Detection

Object detection is utilized for recognizing objective items. It is a strategy of PC vision which is organized in recordings or pictures. Item recognition calculations are shown in outcome when a client is introduced as a picture or video and is approached to discover an article. Hence at that point the client is fit for playing out that task promptly. The article recognition's primary design is to recover this shrewdly using a PC by Object identification is a fundamental innovation behind cutting edge driver help frameworks where the vehicle driving paths are distinguished or used to perform walker location to upgrade street wellbeing. These days Object identification is generally utilized in video reconnaissance or picture recovery frameworks fields.

6.1.2 Deep Learning

Artificial insight comprises of deep learning capacity that mimics the human cerebrum and information preparing and creating designs for dynamic and deep learning is a subset procedure of AI in man-made Artificial Intelligence (AI) that has learning unaided organizations ability of information that isn't orchestrated or marked. Profound learning helps a progressive degree of fake neural organizations to complete the methodology of AI. The fake neural organizations are constructed like human cerebrums associated together like a web portraying neuron hubs. The varieties among customary projects and profound learning is that customary projects investigating the liner information while chipping away at profound learning frameworks measure information with a computational methodology utilizing its different leveled capacities.

6.1.3 Geometric Depth Modeling

The two methodologies are obtained from the mathematical planning capacities where both the frameworks are adjusted to a specific profundity signal component that is changed from the estimation. Geometrics has same demonstrating capacities have been applied in before works, for instance wedgelet and plane models or form model.

6.1.3.1 Plane fitting

The essential guideline of this profundity signal demonstrating approach is approximating the sign of a rectangular square by a direct model that depicts a plane. This sort of model focuses on a nearby estimation of profundity blocks with a planar sign trademark-normally introducing level scene territories or articles, With the example esteems dM(u, v) of the plane model of a profundity block is characterized by a straight capacity as follows:

$$d_M(u, v) = d_o + m_u \cdot u + m_v \cdot v \tag{6.1}$$

As do is utilized for characterizing the counterbalance at position (0, 0) and m_(u/v) the incline of the plane in both synchronized ways as given in a profundity block with native common qualities d(u, v), bringing about best estimate by a plane model for recognizing the plane with boundaries do that utilizes the couple of bending grouped with the native sign which is an overall methodology for finishing of less contortion of direct model for introductory arrangement of test esteems is called as straight relapse. The many times it utilizes the mutilation metric for this is Mean Squared Error (MSE), so the least squares direct relapse technique infers the straight model with the least MSE. For test esteems appointed with more than one facilitate, the technique is reached out to different straight relapse. If there should arise an occurrence of two directions, similar to the (u, v) of our profundity block, this is additionally alluded to as least-squares relapse plane or plane fitting.

6.1.4 Depth Coding Based on Geometric Primitives

The compelling profundity pressure utilizes calculation based profundity displaying approach which comes first of its quality to intently address prescient coding of the sign of a profundity block. In like manner, the data or boundaries disclosing the model needed to be available at the decoder remaking. On a basic level the vital information either gathered from accessible wellsprings of recently decoded pictures and squares (forecast) or assurance at the encoder and changed in the piece stream (expected)-typically joined such that the variety among expected and assessed data is communicated at the encoder and closed basing on the stretch out which is expected based on the data is sent to the decoder and large dependent on an expense work that adjusts the trade among rate and mutilate ion, alluded to as rate twisting improvement. These segments gives an outline, forecast, and flagging techniques that are required for executing our profundity signal demonstrating in a coding structure called MVV (Multi View Video).

6.2 Video Analytics Design Using Depth Intra Coding

Flow graph of video analytics design shown in Figure 6.1, using depth intra coding. Pre analysis of coding unit is done for given input. After pre analysis, the features are selected from the obtained data. The fast depth intra coding

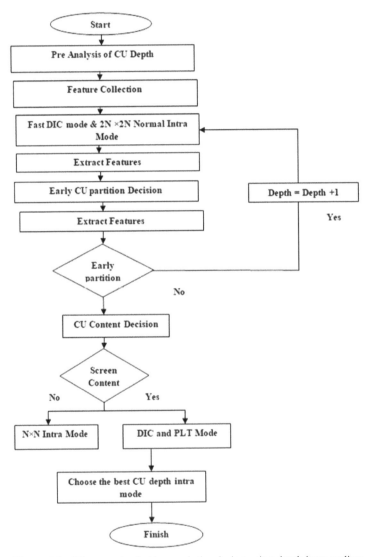

Figure 6.1 Flow graph of video analytics design using depth intra coding.

mode and 2N × 2N intra mode will perform the operation. If the depth size is not equal, then it will again perform its operation from the pre analysis. If the partition is equal then it will depend on the increment of depth and if it is not equal then C content decision will take decision and further process to the screen content. If screen content is not equal then it will perform the N × N intra mode and if it is equal then it performs DIC and PLT mode. At last best coding unit depth intra mode is chosen.

One can track down that enormous coding unit s are bound to be picked as ideal coding unit for "level" or "foundation" locals, where expectation residuals will in general be little and further split for the most part brings little forecast improvement yet expanded side data [9]. For areas which contain last or edges, little coding units are bound to be picked as ideal coding unit size which is expected to be huge. Coding units under the Intra 2N × 2N mode alongside the coding unit split data, where the early partitions address that current coding units should be part and the red spots show that the current coding units are ideal and will not be part further. It tends to be seen that coding units with little residuals like to be coded with enormous coding unit size, which consists of the huge residuals are bound to be part further for more exact expectation. This is particularly valid for coding units at profundity 0. In view of the above perception, this propose to anticipate the coding unit profundity dependent on remaining data where coding unit parting measure will be done.

6.3 Results

Comparison tabular form of existed and proposed system. In this BDBR rate, time saving, signal to noise ratio, complexity and accuracy are given in detail manner. BDBR is decreased in proposed depth intra coding system compared to normal intra coding [10]. Complexity is also decreased in proposed system. Signal to noise ratio is reduced and time saving is increased.

Table 6.1 Comparison of parameters

S. No	Parameters	Normal Intra Coding	Depth Intra Coding
1	Bjonteggard Delta Bit Rate (BDBR)	3.05	2.72
2	Time saving	36.70	48.89
3	Signal to noise ration	More	less
4	Complexity	High	low
5	Accuracy	High	Low

84 *Machine Learning-Based Intelligent Video Analytics Design*

Figure 6.2 compares the complexity, accuracy, and signal to noise ratio of normal intra coding and depth intra coding system for video analytics.

Figure 6.3 compares Bjonteggard Delta Bit Rate (BDBR) of normal intra coding and depth intra coding system for video analytics. Compared to normal intra coding, depth intra coding system will reduce effectively.

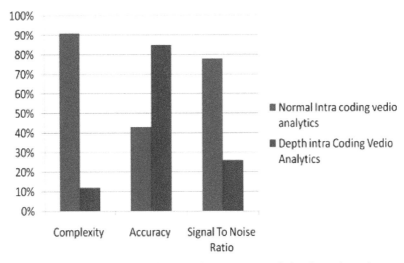

Figure 6.2 Comparison of complexity, accuracy, and signal to noise ratio.

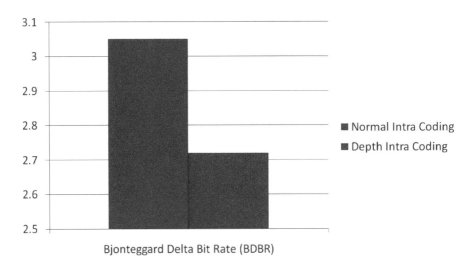

Figure 6.3 Comparison of BJONTEGGARD Delta Bit Rate (BDBR).

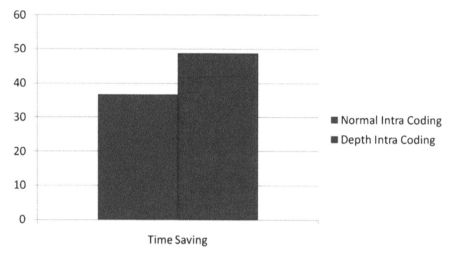

Figure 6.4 Comparison of time saving.

Figure 6.4 compares time savings of normal intra coding and depth intra coding system for video analytics. Compared to normal intra coding, depth intra coding system will save the time very effectively.

6.4 Conclusion

Hence, in this project a novel machine learning based intelligent video analytics design using depth intra coding was implemented. Best CU depth intra mode is chosen at last to get effective output. From results, it can conclude that it gives effective results in terms of complexity, accuracy, and signal to noise ratio, BDBR rate, and time saving. The proposed video analytics using depth intra coding system will improve the efficiency in very efficient way.

References

[1] Multi-scale object detection in remote sensing imagery with convolutional neural networks – Scientific Figure on Research Gate. Available from: https://www.researchgate.Net/figure/The-architecture-of-Faster-RCNN_fig2_324903264 [accessed 15 Feb, 2020].

[2] Banu, Virgil Claudia, Elena Mădălina Costea, Florin Codrut Nemtanu, and Iuliana Bădescu. "Intelligent video surveillance system." In 2017

IEEE 23rd International Symposium for Design and Technology in Electronic Packaging (SIITME), pp. 208–212. IEEE, 2017.
[3] P. Ji, Y. Kim, Y. Yang, Y.-S. Kim, "Face occlusion detection using skin color ratio and LBP features for intelligent video surveillance systems", Proc. Federated Conf. Comput. Sci. Inf. Syst., pp. 253–259, 2016.
[4] G. Sullivan, J. Boyce, Y. Chen, J.-R. Ohm, C. Segall, and A. Vetro, "Standardized extensions of high efficiency video coding (HEVC)," Selected Topics in Signal Processing, IEEE Journal of, vol. 7, no. 6, pp. 1001–1016, Dec. 2013.
[5] K. Muller, P. Merkle, and T. Wiegand, "3-D video representation using depth maps," Proceedings of the IEEE, vol. 99, no. 4, pp. 643–656, april 2011.
[6] P. Merkle, Y. Morvan, A. Smolic, D. Farin, K. Muller, P. H. N. de With, and T. Wiegand, "The effects of multiview depth video compression on multiview rendering," Image Commun., vol. 24, no. 1-2, pp. 73–88, Jan. 2009.
[7] Y. Morvan, P. H. N. de With, and D. Farin, "Platelet-based coding of depth maps for the transmission of multiview images," in Proc. SPIE, vol. 6055, 2006, pp. 60 550K–60 550K–12.
[8] H. Oh and Y.-S. Ho, "H.264-based depth map sequence coding using motion information of corresponding texture video," in Advances in Image and Video Technology, ser. Lecture Notes in Computer Science, L.-W. Chang and W.-N. Lie, Eds. Springer Berlin Heidelberg, 2006, vol. 4319, pp. 898–907.
[9] S. Weisberg, Applied Linear Regression, ser. Wiley Series in Probability and Statistics. John Wiley & Sons, Inc., 2005.
[10] G. Bjontegaard, Calculation of Average PSNR Differences Between RD Curves (VCEG-M33), document ITU-T SG16 Q.6, VCEG Meeting, Austin, TX, USA, Apr. 2001.

7

A Novel Approach for Automatic Brain Tumor Detection Using Machine Learning Algorithms

G. Sindhu Madhuri[*], T. R. Mahesh, and V. Vivek

Faculty of Engineering and Technology, Department of Computer Science and Engineering, JAIN (Deemed to be University), Bangalore, India
E-mail: g.sindhumadhuri@jainuniversity.ac.in;
t.mahesh@jainuniversity.ac.in, v.vullikanti@jainuniversity.ac.in
[*]Corresponding Author

Abstract

Brain tumor is an abnormal tissue growth inside the human skull. In the health care sector, many doctors and researchers are examining for the early prediction of brain tumor disease which leads to many risk factors like brain cancer, brain diseases that can be identified in brain tumor surveillance decision support systems. One of the most popular ways for detection of brain tumor is by analyzing the important information about abnormal tissues that are present in Magnetic Resonance Images (MRIs). As there are lots of improvement and research growth in clinical diagnosis to perform, monitor, and analyze many complex tasks in various fields of medical imaging using machine learning algorithms and medical robotic imaging using deep learning algorithms. However, automatic detection of brain tumor using machine learning algorithms and its early detection of risk factors using nano-robotic health care systems is a challenging and novel approach. Automatic detection of brain tumor and its risks using nano-robotics will focus more on present art of the technology. This chapter highlights the importance of machine learning algorithms toward nano-robotic systems that will provide high statistical measures like accuracy, sensitivity, precision, etc., for the real-time monitoring with nano-robotics by using mobile phones, satellites, sensors, etc. It benefits in minimizing the potential risks connected to human health or environment toward nanotechnologies. This novel approach is highly efficient and effective for future applications by enhancing online monitoring automatic detection of brain tumor nano-robotic systems.

Keywords: Medical Imaging, Brain Tumor Detection, Machine Learning, Nano-Robotic Systems.

7.1 Introduction

There are lots of researches in the medical field and many individual human beings are suffering from brain diseases like brain tumor from past decades [1]. More than a million humans between young and old age ranging between 18 years and 60 years are suffering from many risk factors due to brain cancer, brain tumor, and other brain diseases like artery thickening, cancer prognosis, etc. [2]. Brain tumor is an abnormal tissue growth inside the human skull. The human skull space is very less and the tumor growth inside skull could cause symptoms like enema due to intracranial pressure, displacement, less blood flow, and infects abnormal cells, other soft tissues that control the daily functioning of human body will spread to our brain. This will originate and alert every individual tumor growth inside the skull based on different features like tumor size, type of the tumor, location, structure, etc. Brain tumors are classified as primary brain tumor and secondary brain tumor, where primary brain tumor growth detected in human brain can be benign (means non-cancerous) or malignant (means cancerous) and secondary brain tumor growth detected in human brain can be malignant as shown below in Figure 7.1. Secondary brain tumor growth is dangerous which can lead to serious risk factors w.r.t age, exposure to chemicals, exposure to radiations which can increase the death rate in humans if it is not detected in the early stage [3].

Primary brain tumors are developed from brain cells, nerve cells, glands, and membranes that are surrounded in human brain called meninges/meningiomas. Meningiomas occur commonly in women than men between the ages of 40 and 70. Secondary brain tumors begin from one part of the human body and metastasize to the human brain. Few symptoms that occur in parts of human body like kidney cancers, lung cancers, and skin cancers can slowly metastasize to the brain. Brain tumor diagnosis is one of the complicated processes which involve many specialist opinions by visualizing and analyzing brain scans to acquire better treatment. Brain scan is obtained

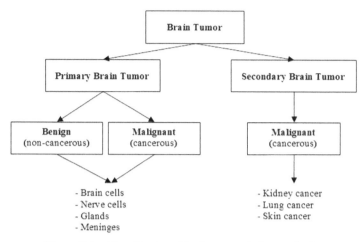

Figure 7.1 Classification of brain tumors in human brain.

from Magnetic Resonance Image. MRI is one of the better primitive and most popular diagnostic ways to detect the different features based on various risk factors of tumor inside our human body. It basically gives us quality images for the detection of different categories of tumor in the brain image than computed tomography image for further treatment and diagnosis [4].

7.1.1 Medical Imaging

Medical imaging from the past 20th century is one of the significant techniques that are mainly applicable to diagnose the captured medical images for treatment purposes and to provide quantitative, functional, structural, and anatomical measurements of the tissues. It originally started with X-rays and introduced computer-based image analysis in the 1960's and taken a big step with the introduction of digital imaging in the 1970's. Medical imaging is not very simple to understand and introduced many reasons for analyzing medical images like clinical study, diagnosis support, treatment, and surgery. To analyze the treatment options to many patients for diagnosing any medical images, researchers introduced many imaging techniques and various modalities. Analyzing and processing of medical images can be carried out by clinicians, radiologists, engineers, and researchers for better understanding of the anatomy of every individual patient. Different medical imaging and segmentation modalities are used for the detection of tumor tissues that are infected in our human brain. MRI uses magnetic waves and radio waves mechanism to produce accurate sufficient information and measurements of tissues and structure of brain cells.

MRI images that capture human brain are digital images that are stored into computer system for further analysis and studying human anatomy. But the MRI images consist of noise that is caused due to its performance [5]. For noise removal of the captured MRI images, we analyze or undergo any medical diagnosis, decision support systems used, computers connected to many tools and develop many methods for detecting the type of tumor or for extracting the features of tumor. Sometimes, MRI brain images may give us clear appearance of tumor or sometimes may not be clear where physicians or doctors may also face difficulty to quantify the area of tumor. So, the computers are connected to many tools, and development of many methods in the healthcare sectors is carried out with the help of many image processing techniques, machine learning algorithms toward nano-robotic systems to diagnose or quantify the tumor area and detect other risk factors [6, 7].

7.2 Image Processing Approach-Detection of Brain Tumor From Mri Images

There is lot of research growth using image processing techniques for the detection of tumor and getting exact outlines of tumor. Basically, medical images are acquired with various electromagnetic spectrum bands and also various sensors are used for acquisition of images that are suitable for a particular purpose. Image processing techniques are mainly used for characterizing the brain tumor, detection of tumor location, analyzing the tumor, etc. Based on size, shape, along with location, and appearance, presence of brain tumor can vary from one patient to another patient, also detection and analysis is a tedious process for medical diagnosis in healthcare sectors. Researchers started working on resulting medical images by describing the existing and multi-modal digital imaging techniques for the detection of brain tumor. To recognize every stage of benign or malignant tumor, image processing techniques are applied by taking input MRI images [8]. The framework of identifying brain tumor using various image processing techniques is shown in Figure 7.2.

The steps for detection of brain tumor using image processing techniques are as follows:

Step 1: Input MRI image—In this step, brain MRI image is taken as input image for further processing for identification of tumor. Images are classified

7.2 Image Processing Approach-Detection of Brain Tumor From Mri Images

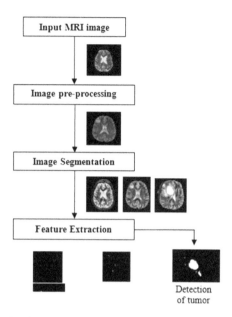

Figure 7.2 Brain tumor detection system-Image Processing Approach.

as MRI image-normal images and MRI image-tumor images (as shown in Figure 7.3, below) and can be taken as input images for further analysis and processing.

Step 2: Image Preprocessing—In this step, the preprocessing is carried out to the input MRI image for better enhancement of the image. It converts input image into grayscale image and performs contrast enhancement as shown in Figure 7.4 below

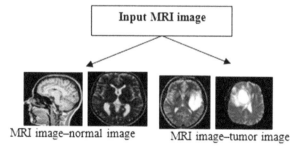

Figure 7.3 MRI input images.

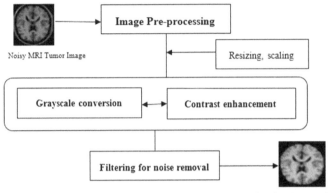

Figure 7.4 Image preprocessing stage.

In this stage, the image resizing, scaling, and alignment are carried out using geometrical transformations like rotation, translation, scaling, and resizing for better enhancement of input images. As the captured MRI images contain noise, the noise pixels in images can be filtered by considering the neighbor pixels. Based on the type of noise distribution model, many filtering algorithms are applied to the input images or detect features like edges, corners, and so on. Filtering techniques like mean filter, median filter, average filter, gaussian filter, etc. can be used to enhance the MRI input images. Few features like edges, corners can't be detected successfully sometimes due to high noise levels [9, 10]. While the process of removing noise of input image, filtering techniques will keep hold of the features and pixel values. And also all the affected pixels and unaffected pixel will be used to calculate mean, average, to replace the resultant pixel value for further analysis. The output in this pre-processing is getting free noise from input MRI image.

Step 3: Image Segmentation—In this step, the input image is basically divided into regions by considering the same attribute. The aim of this stage is to extract features of image by using segmentation techniques like thresholding, clustering, etc. Mainly thresholding segmentation technique is the commonly used technique which converts the grayscale pre-processed image to binary image. This thresholding technique is also called as intensity-based regimentation technique to extract the object from its background. Segmented image with dark background and light tumor area is achieved as output in this step [11, 12]. The process is shown below in Figure 7.5.

7.2 Image Processing Approach-Detection of Brain Tumor From Mri Images

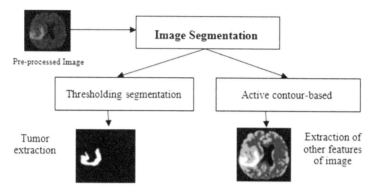

Figure 7.5 Image Segmentation stage.

In Figure 7.5, active contour-based process is carried out for performing image segmentation and tracking the boundaries of the image. Boundary tracking is done with initial shapes that are present in the form of contours, curves, etc. Main advantage of using this is to perform segmentation by partitioning the images based on regions along with the continuous boundary [13, 14].

Step 4: Feature Extraction—As shown in Figure 7.6, it is a method to extract important features like region of interest (ROI), extract tissues from maximal area from the previous stage segmented image by considering minimal elements for representation of dimensionality. The main task in

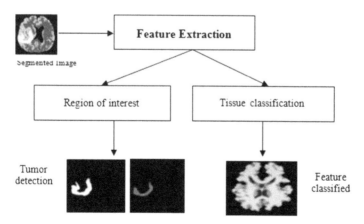

Figure 7.6 Feature extraction stage.

this step is to extract the useful information for detection of brain tumor. Tissue classification is carried out for the classification of MRI image–normal image, MRI image–brain image (i.e., abnormal image) by using various classification techniques like k-means, support vector machine (SVM), etc.

One of the challenging tasks in image processing approach is segmenting of tumors from MRI brain images. Based on pixel-level and image-level observations, segmentation process is carried out. And using feature exaction techniques for extracting the features gives us the detection of tumor portion for measuring the further evaluation metrics like error, accuracy, etc. But using image processing approach, prediction of diseases and better diagnosis due to tumor size in brain image may face difficulty, as the large quantity of hidden information is stored in health care sectors with few risk factors. This analyzing may extend for achievement of better results for early prediction of brain tumor using machine learning approach as discussed in next section.

7.3 Machine Learning Approach-Detection of Brain Tumor From MRI Images

Based on risk factors, researchers examined surveillance machine learning approach techniques for the detection of brain tumor. Using machine learning approach, the detection of tumor assures few important parameters like high efficiency, better accurate results compared to image processing approach. There are many automated segmentation and classifications of machine learning techniques that are proposed for extraction of tumor from MRI brain images datasets. Based on various input brain data sets, identification of benign and malign type of tumor diseases is carried out in this machine learning approach [15]. The process for the detection of brain tumor using machine learning approach is carried out in various stages as show in Figure 7.7 below.

The steps for detection of brain tumor using machine learning techniques are as follows:

Step 1: Input MRI image datasets—In this step, brain MRI image datasets are taken as input image for further processing in the identification of tumor. The datasets consist of two types of images, benign and malignant tumor images. Based on the tumor detection of output image, this approach detects whether the type of disease is benign (non-cancerous cells) or malignant (Cancerous cells). Figure 7.8 shows MRI brain datasets.

7.3 Machine Learning Approach-Detection of Brain Tumor From MRI Images

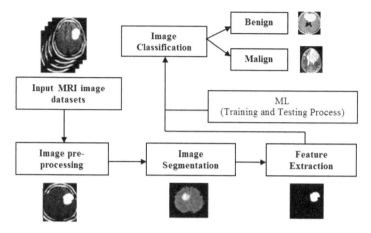

Figure 7.7 Brain tumor detection system-Machine Learning Approach.

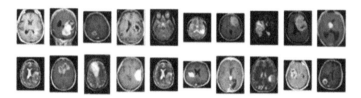

Figure 7.8 MRI brain dataset.

Step 2: Image pre-processing—Image pre-processing step is the important task to improve the quality of MRI input images, so that it makes the image in the form best suited for further processing. It helps to improve few important parameters like visual appearance, peak signal-to-noise ratio, improves noisy images by preserving the clarity of its edges, detect the outliers, etc. To enhance the improvement of visualization, noise, and peak signal-to-noise ratio parameters in input images, converts input image from RGB-to-gray using binarization method, Adaptive Contrast Enhancement method, gaussian methods, filtering methods are applied [16, 17]. The process for the image pre-processing in machine learning approach is shown in Figure 7.9 below.

Step 3: Image Segmentation—Segmentation is the important process which separates the tumor from pre-processed brain tissues. It is the common technique in image processing and in machine learning approach. It analyzes useful information for further diagnosis and planning better treatment based on the type of tumor disease. Segmentation is mainly used to quantify the

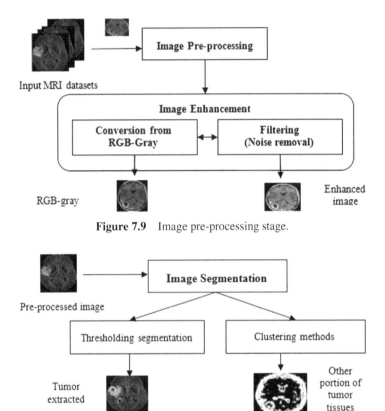

Figure 7.9 Image pre-processing stage.

Figure 7.10 Image segmentation stage.

tumor tissues, classify the tumor features into white pixels, gray pixels, and to segment the image. Researchers developed many segmentation methods like Otsu's binarization method, k-means, c-means, different clustering methods, etc., for extracting the tumor segment [18–20]. The process of image segmentation is shown in Figure 7.10:

Step 4: Feature Extraction—In this step, feature extraction takes place by considering the image-based features and its coordinate-based features present in input MRI image. Based on input image data, image-based feature extraction process is carried out by considering intensity features, histogram-based features, and texture- and shape-based features. And the coordinate-based or texture-based feature extraction process is carried out based on spatial structures, coordinate feature or extracting texture tissue types in the same coordinate system. Many feature extraction techniques like wavelength

7.3 Machine Learning Approach-Detection of Brain Tumor From MRI Images

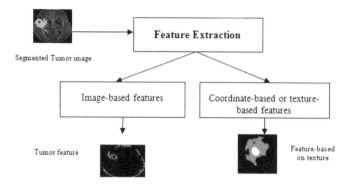

Figure 7.11 Feature extraction stage.

transformation algorithms, random forest algorithm, SVM algorithm etc., can be applied in this stage [21–23]. Figure 7.11 shows the feature extraction process.

Step 5: Image Classification–Basically, machine learning algorithms are used to classify the MRI brain images either as benign (normal-non-cancerous cells), malign (abnormal-cancerous cells). Machine learning algorithms learn and classify to make intelligent decisions based on the outcome of MRI tumor image. The process of image classification is shown below in Figure 7.12.

This image classification is carried out based on classifier and feature classification mainly analyzing classification machine learning model by

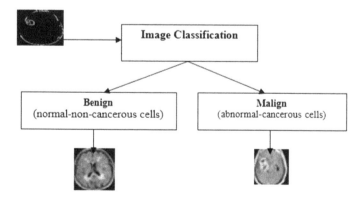

Figure 7.12 Image classification stage.

performing training and testing process. The training set is used during the process of training model for adjustment of hyper-parameters. And the test set is one of the small sets used to check the performance evaluation for final model on some sample images. Feature classification process is extracted to refine the feature set for achieving high classification accuracy. Classifier training feature is carried out for extracting the normal and abnormal type of images to perform further diagnosis and better treatment [24, 25]. Classification algorithm convolution neural network machine learning algorithm is proposed to improve the performance evaluation metrics like error, precision, recall, accuracy, etc. Further diagnosis of brain tumor can be acquired based on outcome values: True Negative (TN)-which indicates no brain tumor, i.e., benign (normal-non-cancerous cells), and False Negative (FN)-indicates no prediction of brain tumor in patients, True positive (TP)-indicates the prediction of brain tumor in patients, and False positive (FP)-indicates no brain tumor but predicts patient having brain tumor. But researchers found lack of automatic detection of brain tumors based on risk factors by using machine learning approach. So, the research is been extended using an novel approach using nano-robotics for early and automatic detection of brain tumor to save patient lives as discussed in next chapter.

7.4 Nano-Robotic Approach-Detection of Brain Tumor From Mri Images

Nano-robotics has good potential for advances in medical diagnosis and guide treatment to patients. For the automatic detection of brain tumor using machine learning algorithms and its early detection of risk factors, a new approach is developed by using nano-robotic health care systems as shown in Figure 7.13. It is one of the challenging and novel approaches at present.

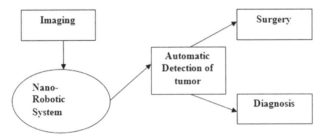

Figure 7.13 Nano-robotic system-A novel approach for automatic detection of brain tumor.

Automatic detection of brain tumor and its risks using nano-robotics will focus more on present art of the technology. It benefits in minimizing the potential risks connected to human health or environment toward nanotechnologies. This novel approach is highly efficient and effective for future applications by enhancing online monitoring automatic detection of brain tumor nano-robotic systems. This approach takes input as MRI brain image for the automatic detection of brain tumor. Nano-robotic system acts as a precision machine for the real-time monitoring with nano-robotics by using mobile phones, satellites, sensors, etc. This system monitors and analyzes the input image and detects the type of tumor automatically and its tumor disease in the early stage. Mainly by highlighting the importance of machine learning algorithms toward nano-robotic systems, it will provide high statistical measures like accuracy, sensitivity, precision, etc., for undergoing better diagnosis or surgery. In the overview of most recent developments, research opportunities project lots of impact on human health.

References

[1] Logeswari, T. and Karman, M. "An Improved Implementation of Brain Tumor Detection Using Segmentation Based on Soft Computing", Journal of Cancer Research and Experimental Oncology, 2, 6–14, 2010.

[2] Abd El Kader Isselmou, Shuai Zhang, Guizhi Xu. "A Novel Approach for Brain Tumor Detection Using MRI Images", J. Biomedical Science and Engineering, 9, 44–52, 2016.

[3] N. Singh and Naveen Chodhary, "A review of brain tumor segmentation and detection techniques through MRI", International Journal of Computer Application, 2014, vol. 103(7): 8975–8887, 2014.

[4] Ms. Swati Jayade, Dr. D. T. Ingole, Prof. Mrs Manik D. Ingole, "Review of Brain Tumor Detection Concept using MRI images", International Converence on Innovative Trends and advances in Engineering and Technology, ISBN-978-1-7281-1901-4, 2019.

[5] American Cancer Society. Cancer Facts & Figures 2016. Atlanta, Ga; 2016.

[6] B. Shoban Babu, S. Varadarajan, "A Novel Approach To Brain Tumor Detection", International Journal Of Engineering Sciences & Research Technology, 6(7), DOI: 10.5281/zenodo.834556, 2017.

[7] K. T. Keerthana and S. Xavier, "An intelligent system for early assessment and classification of brain tumor". in Proc. 2nd Int. Conf.

Inventive Commun. Comput. Technol. (ICICCT), Coimbatore, India, pp. 1265–1268, Apr. 2018.
[8] K. Elavarasi, A. K. Jayanthy. "Soft sensor based brain tumor detection using CT-MRI" International Journal of Science, Engineering and Technology Research (IJSETR) Volume 2, Issue 10, October 2013.
[9] Jaya, K. Thanushkodi, M. Karnan, Tracking Algorithm for De-Noising of MR Brain Images, International Journal of Computer Science and Network Security, 9(11), November 2009, 262–267.
[10] B. Goossens, A. Pizurica and W. Philips, "Image denoising using mixtures of projected Gaussian scale mixtures", IEEE Transactions on Image Processing, vol. 1.18, no. 8, pp. 1689–1702, 2009.
[11] R. Saini, M. Dutta, "Image segmentation for uneven lighting images using adaptive thresholding and dynamic window based on incremental window growing approach", Int. J. Comput. Appl, 56 (13) (2012), pp. 31–36.
[12] Ehab F. Badran, Esraa Galal Mahmoud, and Nadder Hamdy, "An Algorithm for Detecting Brain Tumors in MRI Images" IEEE Conference Publication Annual IEEE, Publication year 2010.
[13] M. Kass, A. Witkin, D. Terzopoulos, "Snakes, active contour model", Int J Comput Vision, 1 (4) (1988), pp. 321–331.
[14] Sivaramakrishnan, Dr. M. Karnan, "A Novel Based Approach for Extraction of Brain Tumor in MRI Images Using Soft Computing Techniques", International Journal of Advanced Research in Computer and Communication Engineering Vol. 2, Issue 4, 2013.
[15] Nelly Gordillo, Eduard Montseny, Pilar Sobrevilla, "State of the art survey on MRI brain tumor segmentation", Magnetic Resonance Imaging 31 (2013) 1426–1438.
[16] A. Demirhan, M. Toru, and I. Guler, "Segmentation of tumor and edema along with healthy tissues of brain using wavelets and neural networks," IEEE Journal of Biomedical and Health Informatics, vol. 19, no. 4, pp. 1451–1458, 2015.
[17] S. Lal and M. Chandra, "Efficient algorithm for contrast enhancement of natural images," International Arab Journal of Information Technology, vol. 11, no. 1, pp. 95–102, 2014.
[18] Prachi gadpayleand, p. s. Mahajani, "detection and classification of brain tumor in MRI images", international journal of emerging trends in electrical and electronics, ijetee — issn: 2320–9569, vol. 5, issue. 1, july-2013.

[19] Bakes s., Reyes m., menze, b.: identifying the best machine learning algorithms for brain tumor segmentation, progression assessment, and overall survival prediction in the Brats challenge. In: arxiv:1811.02629 (2018).
[20] Prastawa m., Bullitt e., Greig, g., 2004. A brain tumor segmentation framework based on outlier detection. Medical image analysis 8, 275–283.
[21] Bauer s., Nolte, Reyes m., 2011. Fully automatic segmentation of brain tumor images using support vector machine classification in combination with hierarchical conditional random field regularization, in miccai, pp. 354–361.
[22] Gotz m., Weber c., Blocher j., Stieltjes b., Meinzer, h. p., Maier-Hein, k., 2014. Extremely randomized trees-based brain tumor segmentation, in proc of brats challenge - miccai.
[23] Prachi gadpayleand, p. s. Mahajani, "detection and classification of brain tumor in MRI images", international journal of emerging trends in electrical and electronics, ijetee — issn: 2320–9569, vol. 5, issue. 1, july-2013.
[24] N. Varuna Shree, T. N. R. Kumar "Identification and classification of brain tumor MRI images with feature extraction using DWT and probabilistic neural network", © Springer, Brain Informatics, pp. 23–30.
[25] J. Seetha and S. Selvakumar Raja "Brain Tumor Classification Using Convolutional Neural Networks", Biomedical & Pharmacology Journal, 2018. Vol. 11(3), p. 1457–1461.

8

A Swarm-Based Feature Extraction and Weight Optimization in Neural Network for Classification on Speaker Recognition

G. Raja[1], P. Salini[2], M. Pradeep[3], and Terrance Frederick Fernandez[4]

[1]Research Scholar, Department of Computer Science & Engineering,
Puducherry Engineering College, Puducherry, India
[2]Assistant Professor, Department of Computer Science & Engineering,
Puducherry Engineering College, Puducherry, India
[3]Assistant Professor, Department of Information Technology, Rajiv Gandhi College of Engineering and Technology, Puducherry, India
[4]Associate Professor, Department of Information Technology,
Dhanalakshmi Srinivasan College of Engineering & Technology,
Tamil Nadu, India
E-mail: raja@pec.edu; salini@pec.edu; mpradeepcseb@yahoo.com; frederick@pec.edu

Abstract

Speaker recognition is being evolving for the past decade in a noticeable way. Every person who is using mobile phones is exercising this technique. Mel-frequency cepstral coefficient (MFCC) and Linear prediction cepstral coefficient (LPCC) are the two most popular feature sets used for speech signal analysis. Nowadays, deep learning has become a part of every material which is developed using AI. By using this technique, we can be more successful in finding out the speakers in variable environments. Our chapter represents a speaker recognition method inspired upon MFCC and deep neural networking. Usually, when there is more number of speakers, rate of recognition is kept at the rock bottom. The problems and its solutions regarding participation of more number of speakers are discussed here. By giving more training samples to system, the accuracy of finding the speakers will be bettered through our proposed system.

Keywords: Mel-Frequency, Deep Neural Networks, Artificial Intelligence, Speaker Recognition.

8.1 Introduction

Speaker recognition is a hot topic under Ambient Intelligence (AmI) which has multiple phases for research [1]. Voice recognition provides a non-obtrusive human-computer interaction. Voice recognition can be categorized as speaker recognition, that sorts out a person's voice and speech recognition that sorts out the words. Speech recognition technologies are widely employed in human–computer interaction. On the flipside, speaker recognition is widely employed for authentication.

Feature Extraction and Classification are the two major factors that distinguish a speaker from others where the voice is treated as spectral which further recognized as bands for efficient classification. ***Feature Extraction*** is a sensible task since a wrong choice of feature from a dataset may increase the probability of False Positive and it comes under optimization since the possibility of choosing an appropriate feature is higher in dimension as well as in numbers. ***Classification*** of speaker based on the extracted feature is another challenging task where the classification needs to be more accurate. Though proper feature extraction techniques exist, if the classification methodology goes wrong then the feature extraction will also be stagnated. In order to address these factors, we propose our objectives to study and design a better feature extraction and classification algorithm for effective speaker recognition.

8.1.1 Swarm-based Feature Extraction Merits

- Swarm-based feature extraction can bring more accuracy rates when compared to previous techniques.
- Nature-inspired evolutionary algorithms will be more efficient while used in speaker recognition in human beings.
- The convolutional neural networks provide a deeper idea in extracting the features of individual speakers.

- These swarm-based algorithms are not still implemented in speaker recognition techniques and thus good for research.
- Mel-frequency cepstral coefficient (MFCC) feature extraction can be enhanced by using these evolutionary algorithms.
- Optimization of feature extraction and pattern matching can be done more efficiently by swarm-based algorithms for speaker recognition.

8.1.2 Objectives of Our Chapter

The chapter is objected toward a comprehensive study on feature extraction techniques, a neural network model for speaker recognition, and other soft computing methodologies to optimize it. Furthermore, we design and develop a swarm-based algorithm for effective feature extraction from multiple speaker's voice database.

Secondly, we model the proposed algorithm for effective weight optimization on feed forward neural network for classification of data models based on recognized voice. Then, we design and develop a testbed environment in order for the evaluation of the performance of our proposed work. We validate and compare the performance of proposed algorithm over multiple performance assessment criteria with other existing techniques.

8.2 State of Art

The theme of the chapter has been under limelight over the decade and it arrives in varied methodologies. MFCC is a noticeable victorious method as it is modeled upon the human auditory region. A study of the same, presents it to have high success rate of recognition and strong robustness against noise in the lower-frequency regions [1]. On the flipside, higher the frequency regions, it captures speaker characteristics information less effectively. This decade, the Artificial Neural Networks have become popularized. Our chapter presents a speaker recognition method based on MFCC and Back-Propagation Neural Networks. Experimental studies have proven that the recognition rate is successful when the number of questionable speakers is not huge. As the number of speakers rises, the rate of apt recognition falls down. The potential problems and solutions are furnished in this chapter. The quantity of training samples must be greater than the number of network model weights, at least 10 times of the original. When the number of speakers increases, the number of training samples required also spikes up.

8.2.1 Mel Frequency Cepstral Coefficients (MFCC)

MFCCs are mostly used in the speaker and speech recognition applications and it is accounted by authors in [3]. Since the 1980s, works were voracious that led to the development of these features. Works like employing the correct spectral estimation methods, architecting effective filter banks, and selecting chosen ideas render a significant role in the performance and robustness of these systems. Our chapter provides a comprehensive view of MFCC's techniques that are applied in these works. The details such as accuracy, types of environments, the nature of data, and the number of features are investigated and summarized in the table combined with the corresponding key references. Benefits and drawbacks of these MFCC's enhancement techniques will be discussed. This study will possibly render to raising initiatives toward the development of this in characteristics of robustness features, high accuracy, and less complexity.

8.2.2 Swarm Intelligence (SI)

Swarm proved as a significant topic in AI. The name is derived from the cooperative nature of bees and/or ants. The research of swarm intelligence (SI) optimization is adapted by swarm activities by bees or ants led to the development of swarm UAVs technology [4]. There are 11 SI algorithms that are described in [4] which further presented a detailed analysis report that aided combination of SI and multi-UAV task assignments.

8.2.3 Text-independent Speaker Identification

A human auditory pulse carries variety of data. A plethora of methods were proposed during the past 20 years to solve different speech processing problems such as recognition of speech, auditory emotion, or language being spoken which were discussed by authors in [2]. Text-independent Speaker Identification can give characterization and recognize the information about the auditory information. Speaker recognition can be classified as speaker verification and identification. The former identifies a particular person from the voice. The latter one determines who the speaking person is, of which there is no identity claimer. Considering the identification, the task can be subdivided into text-dependent and text-independent identification. The sole difference is that the system knows the text spoken by the person in the former while the system must be able to sort out the speaker from any well-defined sentence of text for text-independent identification.

8.2.4 Voice Activity Detection (VAD)

Principal objective of VAD is to detect a voice for the purpose of helping speech processing to aid the start and end of the auditory impulse and this is depicted in [7]. Rudimentary nature of VAD is to juice out certain parameters of the input signal. It is then equated with a threshold value. The nature is primarily found from the nature of the noise and the target sound. Decision-making is done when the signal is active and started if the value of the test approaches the upper limit value and ends at the zenith. Selection of the apt threshold will answer the victorious nature of VAD, if the signal is active or not.

8.3 Differential Evolution Technique (DE)

Evolutionary algorithms are the methodologies that score for all classes of optimization. Researchers in [3], accounted that these are capable to give feasible solutions for real-life problems. Requirement of more computational time for optimizing is the root cause of the massive acceptance in evolution optimization techniques in comparison to classic methods & analytical ones. The evolutionary ones come in different flavors. Firstly, the differential evolution algorithm is quite recent one. The prime reason of presenting this algorithm floats over the primary error in genetic ones, that is, the local search negativity. The next section presents a comprehensive survey on SI.

8.4 Survey on Swarm Intelligence

A series of approaches were discussed in [1] by the team headed by Yi Wang and Dr. Bob. Firstly, through proper analysis and the foraging process of ant colony, the features of the same can be used to identify the shortest route. A team of experts in [4] headed by Marco Dorigo researched upon this was popularized as Ant Colony Algorithm (ACO).

In the mid-90s, Kennedy and his group of experts eyed trivial rules and swarm nature of bird feeding, motile nature and coined the Particle Swarm Algorithm (PSO). Further on, the Shuffled Frog Leaping Algorithm (abbreviated as SFLA) considered collections of frog of different intentions and similar among them formed swam. Jumping down from frogs to the pack of wolves, a novel approach called WPA, the Wolf Pack Algorithm, is based on the survival and unity of wolves that are massively profited through pack cooperation.

Brood parasitism is prevalent among birds like cuckoo, which fails in their hatching mechanism but are an expert in this. Researchers in [1] devised Cuckoo Search Algorithm (CS) upon this basis.

Navigators like bats or flying foxes employ echolocation through their ultrasonic pulses from their bodies. This is called as Bat Algorithm (BA). Apart from navigation purposes, glow-worms and fireflies use luminescence in their biology to get a target signal pattern and this is used for communication and researchers in [5] coined Firefly Algorithm (FA) upon the nature of this insect.

Considering insects, once again, the bee colony has a queen surrounded by the drones and the worker ones. Every insect inside the colony has a neat job and has its own unique responsibility. An algorithm is built upon the colony termed as Bee Algorithm. Fruit flies carry an awesome olfactory mechanism within which it aids with an excellent sense of smell and sight. Such an immaculate intelligence is adapted by the researchers in the name of optimization algorithm called as Fruit fly optimization algorithm (FOA).

A massive set of insects and animals are instrumental in transmitting pollen from flowers. This process is carefully studied by researchers in [1] who formed a pollination technique. Finally, an evolutionary approach, named *Eel Algorithm (EA)* was based on the mechanism adapted underwater by fishes like eel, in which, large number of female species evolve into males during starvation times and/or during overpopulation.

The next section of the chapter describes our work with implementation framework.

8.5 Our Framework and Metrics

In our framework, we employed the most common metric which is popularly called Word Error Rate (WER). A particular performance is calculated by comparing a reference transcription with the transcription output by the AI-powered recognizer. Our framework model is furnished in Figure 8.1.

Upon these computed output, we can sort out the number of bugs, which could typically correspond to these classifications that could fall under a particular condition set:

1. Suppose, if an output of the ASR renders a word but it is not found at the reference, then it is classified under Insertions error class (IEC).
2. In another instance, if a word is not found in the target output, then it is classified under Deletions error class (DEC).
3. Finally, if a word is altogether chaosed with another then it is classified under Substitutions error class (SEC).

8.5 Our Framework and Metrics

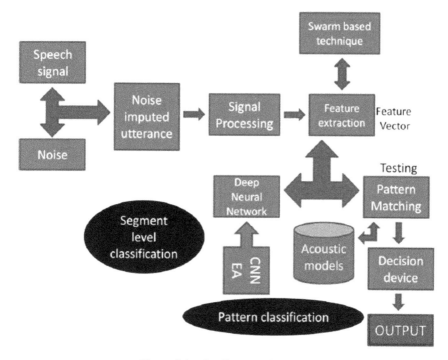

Figure 8.1 Our Framework model.

With the above classes, we can equate WER as:

$$WER = (SEC+DEC+IEC)/n$$

Where, n → words found in the reference

Initially the speaker's speech with noise is given as input. Every speaker will have their own way of modulation and pronunciations of words. As the number of speakers increases, automatically the noise increases, so it needs more signal processing time for extracting the features from the speakers' voices.

In this proposed method, we are using swarm-based technique to extract the features of the noise imputed utterance. After the features are extracted, they are given to the system for training and testing processes. The deep learning methodology makes the system more efficient by adding more hidden layers with the given training samples. By using CNN and evolutionary algorithms, the classification of speakers are performed in the acoustic model. Finally, with pattern matching techniques, the inputted speaker's voice is

110   A Swarm-Based Feature Extraction and Weight Optimization

matched with the correct one inside the system and decision is made. With the help of deep learning when applied to the voice data, recognition of speakers will be more efficient when compared to the existing system.

8.6 Results and Discussion

An implementation was made with a sample set of three speakers with nine audio files coded in mp3 format. The three speakers are pseudonym-ed as S, R, and B. The test set consisted of voices of the speakers in three different moods taken from emotions such as anger, sad, happy, fear, disgust, and surprise. Thus S can take values s_1, s_2, and s_3. Similarly R can take r_1, r_2, and r_3 and B can take b_1, b_2, and b_3. The preprocessing of our work is done by training a voice in combinations of these moods and making the CNN powered system to learn it. There are six samples considered for training set.

The speaker output audio (in mp3 format) is being converted to spectroscopic image with equal 1024×1024 dimensions. The resultant image is again pixilated using image embedding technique in orange data science toolkit. This is done to be fed to the learning algorithms in the data science toolkit. Three algorithms were considered which included k-nearest neighbors (k-NN), Neural Networks, and Random Forest. The confusion matrix of three approaches is given in Figures 8.2–8.4.

		Predicted							
		b1	b2	r1	r1s1	r1s1-cut1	r1s1-cut2	r2	Σ
Actual	b1	1	0	0	0	0	0	0	1
	b2	1	0	0	0	0	0	0	1
	r1	0	1	0	0	0	0	0	1
	r1s1	0	1	0	0	0	0	0	1
	r1s1-cut1	1	0	0	0	0	0	0	1
	r1s1-cut2	1	0	0	0	0	0	0	1
	r2	0	1	0	0	0	0	0	1
	Σ	4	3						7

Figure 8.2   Confusion Matrix for k-NN.

		Predicted							
		b1	b2	r1	r1s1	r1s1-cut1	r1s1-cut2	r2	Σ
Actual	b1	0	0	0	0	0	1	0	1
	b2	0	1	0	0	0	0	0	1
	r1	0	0	1	0	0	0	0	1
	r1s1	0	0	0	0	0	1	1	1
	r1s1-cut1	0	0	0	0	1	0	0	1
	r1s1-cut2	0	0	0	0	0	1	0	1
	r2	0	0	0	0	0	0	1	1
	Σ		1	1		1	2	2	7

Figure 8.3   Confusion Matrix for Random Forest.

8.6 Results and Discussion   111

	Predicted	b1	b2	r1	r1s1	r1s1-cut1	r1s1-cut2	r2	Σ
Actual	b1	1	0	0	0	0	0	0	1
	b2	0	1	0	0	0	0	0	1
	r1	0	0	1	0	0	0	0	1
	r1s1	0	0	0	1	0	0	0	1
	r1s1-cut1	0	0	0	0	1	0	0	1
	r1s1-cut2	0	0	0	0	0	1	0	1
	r2	0	0	0	0	0	0	1	1
	Σ	1	1	1	1	1	1	1	7

Figure 8.4 Confusion Matrix for Neural Networks.

From the Confusion matrix, it is evident that the diagonal values are predominantly 1 for neural networks than other approaches. Figures 8.5 and 8.6 present the test score data that predicted the speaker correctly in terms of probability values. Here too, it is evident to have Neural Network to rightly recognize the speaker.

Scores

Model	AUC	CA	F1	Precision	Recall
kNN	0.6666666666666665	0.142857142857142857	0.0571428571428571	0.0357142857142857	0.142857142857142857
Random Forest	0.9999999999999998	0.7142857142857143	0.619047619047619	0.5714285714285714	0.7142857142857143
Neural Network	0.9999999999999998	1.0	1.0	1.0	1.0

Figure 8.5 Test and score for prediction.

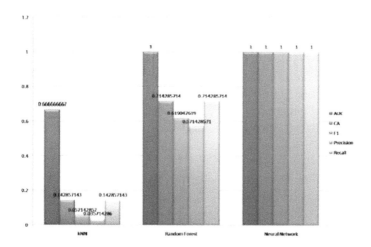

Figure 8.6 Comparison of different AI approaches in terms of prediction.

Thus, we have proved that this proposed framework can extract and classify the voices of multiple speakers in a noisy environment.

References

[1] Yi Wang, Dr. Bob Lawlor "Speaker Recognition Based on MFCC and BP Neural Networks," 2017 IEEE.

[2] Abhinav Anand, Ruggero Donida Labati, Madasu Hanmandluy, Vincenzo Piuri, Fabio Scotti, "Text-Independent Speaker Recognition for Ambient Intelligence Applications by Using Information Set Features", 2017 IEEE.

[3] Mohsen Sadeghi, Hossein Marvi "Optimal MFCC Features Extraction by Differential Evolution Algorithm for Speaker Recognition, 3rd Iranian Conference on Signal Processing and Intelligent Systems (ICSPIS)" 2017 IEEE.

[4] Feng Yang, Pengxiang Wang, Yizhai Zhang, Litao Zheng, Jianchun Lu "Survey of Swarm Intelligence Optimization Algorithms", 2017 IEEE.

[5] Pulkit Verma and Pradip K. Das, "A Comparative Study of Resource Usage for Speaker Recognition Techniques", 2016 IEEE.

[6] Karishma Trinanda Putra, "Voice Verification System Based on Bark Frequency Cepstral Coefficient", Journal of Electrical Technology UMY (JET-UMY), Vol. 1, No. 1, March 2017

[7] Nagwa M. AboElenein1, Khalid M. Amin2, Mina. Ibrahim3, Mohiy M. Hadhoud, IEEE, "Improved Text-independent Speaker Identification System For Real Time Applications", 2016 IEEE.

[8] Sayf a. majeed, Hafizah husain, Salina abdul samad, Ariq f. "Mel frequency cepstral coefficients (mfcc) feature extraction enhancement in the application of speech recognition: a comparison study", Journal of Theoretical and Applied Information Technology, 2015.

[9] Yuan, Yujin, Peihua Zhao, and Qun Zhou. "Research Of Speaker Recognition Based On Combination Of LPCC And MFCC". 2010 IEEE International Conference on Intelligent Computing and Intelligent Systems 3 (2010): 765–767.

[10] Yousra F., Al-Irhaim Enaam Ghanem Saeed, "Arabic word recognition using wavelet neural network", Scientific Conference in Information Technology, November 2010.

[11] Thiruvaran, Tharmarajah, Eliathamby Ambikairajah, and Julien Epps. "FM features for automatic forensic speaker recognition." Interspeech. 2008.

[12] Gonzalez-Rodriguez, Joaquin, et al. "Robust likelihood ratio estimation in Bayesian forensic speaker recognition." Eighth European Conference on Speech Communication and Technology. 2003.
[13] Reynolds, Douglas A. "An overview of automatic speaker recognition technology." Acoustics, speech, and signal processing (ICASSP), 2002 IEEE international conference on. Vol. 4. IEEE, 2002.
[14] Wang, Yue, et al. "Quantification and segmentation of brain tissues from MR images: a probabilistic neural network approach." IEEE transactions on image processing 7.8 (1998): 1165–1181.
[15] Chtioui, Younes, et al. "Comparison of multilayer perceptron and probabilistic neural networks in artificial vision. Application to the discrimination of seeds." Journal of chemometrics 11.2 (1997): 111–129.
[16] Specht, Donald F. "Probabilistic neural networks." Neural networks 3.1 (1990): 109–118.

9

Fault Tolerance-Based Attack Detection Using Ensemble Classifier Machine Learning with IOT Security

A. Arulmurugan[1], R. Kaviarasan[2], and Saiyed Faiayaz Waris[3]

[1] Associate Professor, Department of CSE, Vignan's Foundation for Science, Technology & Research (Deemed to be University), Guntur, Andhra Pradesh, India
[2] Assistant Professor, Department of CSE, RGM College of Engineering and Technology, Nandyal, Andhra Pradesh, India
[3] Assistant Professor, Department of CSE, Vignan's Foundation for Science, Technology & Research (Deemed to be University), Guntur, Andhra Pradesh, India
E-mail: Arulmurugan1982@gmail.com; Kaviarasanr64@pec.edu; Saiyed.cse@gmail.com

Abstract

Computer vision has recently progressed. In our homes and workplaces everywhere, Internet of Things (IoT) devices and applications are being deployed. Deep learning architectures are often fed by continuous data or categorical data collection in these devices. However, since the service provider may make unwanted inferences on the available data, this method poses some privacy and efficiency issues. For simple tasks and lighter models, recent developments in edge processing have paved the way for more effective and private data processing at the source. For larger and more complex versions, however, they remain a challenge. The proposed model is designed between cloud architectures in this paper. The input dataset will be a medical database or sensational dataset; the data has been obtained from IoT. Before this IoT, the cloud structure has been designed with security, and the output of deep learning architecture with ensemble classifiers will have the Intrusion Detection System (IDS) for security. The threshold limit has been set. Similarly, those data beyond the threshold limit will be sensational. Both the cloud architecture will have a similar private key, whether the data is securely transmitted or not. Each information will have its threshold limit, and this has been analyzed; categorized and classified output is obtained from deep learning ensemble classifiers. The experimental results show effective data transmission with higher accuracy when compared with existing techniques.

Keywords: Internet of Things (IoT), Data Processing, Cloud Architectures, Ensemble Classifiers, Intrusion Detection System (IDS), Threshold Limits.

9.1 Introduction

The expanding accessibility of associated IoT gadgets, for example, cell phones and cameras has made them a fundamental and indistinguishable piece of our everyday lives. Most of these gadgets gather different types of information and move it to the cloud to profit by cloud-based information mining administrations like proposal frameworks, directed publicizing, security reconnaissance, wellbeing checking, and metropolitan arranging. Large numbers of these gadgets are financed, and their applications are free, depending on data gathered from the clients' information. This training has a few protection concerns and asset impacts for the clients [1]. For instance, a cloud-based IoT application that offers feeding location administration on the client's camera as its essential errand, may utilize this information for different undertakings like inhabitance investigation, face acknowledgment or scene understanding, which may not be wanted by the client, putting his or her security in danger. IoT is an innovation which is as yet a work in progress and need numerous enhancements in it at an alternate level.

While numerous scientists add one more layer, for example, middleware layer whose capacities are administrating the executives, put away information got from network layer in the data set, and so forth change in the layer doesn't roll out a lot of improvement in IoT innovation just as its imperfections. The point of creating proper climate, originated from the application layer of the IoT. The safeness of IoT is a major test in view of intricacy, heterogeneity, and countless interconnected assets. IoT frameworks can be attacked by altering some hub, getting direct access from its organization or organization that uses it or by injecting a noxious program or by cracking the encryption. In view of these weaknesses, we group the assault into four classes, as actual assault, network assault, programming assault, and encryption assault. From every class, we considered one assault that is generally perilous from all the assaults of that classification. In actual assault, pernicious hub infusion assault has been the risky assault. Since it isn't just halting the administrations yet in addition alter the information. In network assault, sinkhole assault is the most dangerous assault. Intruder can send, change, or drop the packs apart from just pulling in all the rush hour gridlock toward the base station. We can expect worm attack in terms of programming attack. On the web, the most jeopardizing malwares are worms. They reproduce themselves, through loopholes in framework's security and halt the PC. They have the potential to read and change the password, clear the archives, slow the performance of the system and much more [2].

AI frameworks use calculations that improve their yield dependent on experience. Later on, AI will supplant customary enhancement strategies in numerous fields since ML models can extend to incorporate new limitations and contributions without beginning any preparation and they can settle numerically complex conditions. ML models are promptly adjusted to new circumstances, as we are presently seeing with PC frameworks. In the most recent decade, a subset of AI called profound learning (DL) has accumulated a lot of consideration in PC vision and has found new ideal systems for games without the expensive hand-created including the design that is essential in the current era. Profound learning utilizes neural organizations to perform mechanized element extraction from enormous informational collections and afterward utilize these highlights in later strides to characterize input, decide, or create new data. Examination on profound learning for PC vision detonated after the arrival of ImageNet, a curated data set of more than 10 million pictures across 10,000 classifications, which aided in training ML picture characterization models [3].

Flaws inside any organization will undoubtedly occur because of different reasons. An organization called deficiency lenient when it works even regularly when shortcomings happen while the organization is being used. IoT networks are delicate and thusly, should be made shortcoming lenient. IoT networks utilized in the clinical area should be shortcoming lenient as any falsehood stream will cause an overwhelming impact even to the degree of loss of human existence. A little flaw may prompt genuine negative outcomes. At the point when a fault occurs, for the most part, the information obtained is lost. Information should be protected and held at any expense. Utilization of nonvolatile memory inside IoT-based frameworks will help in recuperating from the deficiency of the ordinary activity when a flaw happens. Adaptation to internal failure is fundamental even at the expense of bringing about overhead kick the bucket to utilization of nonunstable recollections. The basic way to deal the problem is to improve the adaptation level in non-critical failure in adding many gadgets in such a way that even if shortfall of one comes there is another gadget to dominate. Processing adaptation to internal failure is as such complex because of the presence of numerous complicated issues. Adaptation to non-critical failure of organization for the most part communicated quantitatively regarding achievement or disappointment rate is the pace of disappointment of highest hubs existing in a Fault Tree—the achievement rate registered as 1—Failure Rate. In a commonplace organization, achievement rate is the likelihood that at any rate one transmission way exists from a communicating gadget to the objective gadget—the disappointment rate got by taking away the achievement rate from 1 [4].

Numerous kinds of flaws occur inside IoT organizations, and every one of the issues should be thought of and discover the strategies to relieve the equivalent. IoT networks commonly are perceived into a few layers. An adaptation to internal failure figuring model utilized could contrast from one layer to another. Adaptation to non-critical failure of an organization for the most part registered utilizing a solitary computational model. A solitary computational model is for the most part not stuffiest as the organizations in each layer may have deterministic or probabilistic conduct. Decision of a geography fitting to the adaptation to internal failure level of the gadgets contained in each layer and decision of the legitimate technique to figure adaptation to non-critical failure of a sub-net will prompt high blame on lenient IoT organization. In this chapter, a composite model that considers distinctive systems administration geographies and adaptation to internal failure registering models that upgrade the adaptation to internal failure level of an IoT network is introduced [5].

9.2 Background

9.2.1 IoT Security Attacks

Various IoT security attack based on each layer are as follows.

9.2.1.1 Perception Layer Attacks

The insight layer comprises of actual articles, for example, sensors and actuators, hubs, and gadgets. An insight layer assault influences the actual article in the IoT framework.

Lack of sleep assault: Absence of rest attack is a sort of attack coordinated on battery controlled sensor centres and devices. Ordinarily, battery energized devices follow a rest routine to expand its lifetime. The absence of rest attack focuses on keeping the centres and devices attentive for a comprehensive time period, which achieves more battery power use and finally shutting down of the centres and contraptions.

9.2.1.2 Network Layer Attacks

The organization layer by and large comprises of organization segments, for example, switches, spans, and different sorts of systems administration parts. An organization layer assault is an assault coordinated toward upsetting the organization segments in the IoT space.

Denial of Service (DoS)/DDoS: DoS is a kind of malevolent assault that points in devouring assets or transfer speed of certified clients. A DDoS is a variation of the DoS which is like the DoS assault however includes different traded off hubs.

Slowloris: The Slowloris is a DDoS attack, where various HyperText Move Convention (HTTP) requests are opened and controlled at the same time between the assailant and target. Slowloris are prepared for failing an application by using irrelevant traffic and aggressors.

Association Time Convention (NTP) Intensification: NTP Enhancement attack is such a reflection-based volumetric DDoS attack where the NTP is abused by the assailant to flood an improved UDP traffic to a host. Thusly, this impacts the host and incorporating establishment causing standard traffic closed off to the resource.

9.2.1.3 Routing Attacks

In directing assaults malignant hubs dispatch steering sorts of assaults to disturb directing activity or for performing DoS assaults.

(a) Sybil Attack: During Sybil assault a malevolent hub breaks the directing framework, and gets to data obstructed by the hub, or the organization gets parceled. This assault is executed by a solitary assailant who makes numerous bogus characters and claims to be various in shared organizations (P-2-P).

(b) Sinkhole Attack: Sinkhole assault is directed by involving a hub which endeavors to draw traffic however much as could be expected from a particular zone, by making itself look interesting to the encompassing hubs dependent on the steering metric. Subsequently, the pernicious hub

draws in all rush hour gridlock from the base station. This at that point gives the assailant to lead further assaults on the framework.
(c) Selective Forwarding Attack: A particular sending assault is fit for directing a DoS assault where malevolent hubs specifically forward parcels. The objective of this assault by and large is to upset steering ways. By and by, it tends to be utilized to channel any convention.
(d) Wormhole Assault: The place of a wormhole attack is to vex the association topography and traffic stream. The wormhole attack happens when a malignant centre tunnels messages among two particular bits of the association through a quick association.
(e) Hello Flood: The invite flood is one of the guideline attacks in the association layer. The invite flood attack enables the aggressor to drive conventional centres to lose power by convincing them to send tremendous howdy groups with high power.
(f) Approval Assaults: Verification-based attacks are used to mishandle the affirmation cycle that is used to check a customer, organization, or application.

9.3 Deep Learning and IoT Security

With the help of notable learning methodologies, we are going to see the best strategies used in IoT security in this section. IoT is used in military nowadays which shows its popularity. Through IoT framework, military insinuates "Web of War zone Things." It is concluded that mostly attacks are carried by mixture of malware in recent survey. To counter-attack the IoT malware, a vector space called Opcodes (Operational Codes) is introduced and the learning approach of Eigenspace is used to organize chivalrous and vindictive application. This approach's sensibility is taken into account opposing the trash code attacks. Relying on four appraisal estimations, specifically precision, exactness, survey, and f-measure the model has been evaluated. Besides, they have considered two other similar examinations subject to the estimation. The achievements of this approach are:

- $99.68\% \rightarrow$ Precision,
- $98.59\% \rightarrow$ exactness,
- $98.37\% \rightarrow$ survey,
- $98.48\% \rightarrow$ f-extent.

An alternative way to protect from trash code attack is also provided by this model. The datasets accustomed to this model evaluation are coincidentally

9.3 Deep Learning and IoT Security

independent. Data's authenticity and quality is effectively refuted. Added to it, the datasets are associated with malware tests. More number of IoT contraptions is now threatened by IoT botnets. To recognize strange traffic in the network of compromised IoT contraptions, harness of DAE is encouraged by the makers of the framework to uproot this risk. To eliminate direct attack, significant learning/steps have been done. Botnets named BASHLITE and Mirai have been used to halt nine IoT contraptions used in business, to analyze their model. Then the model was subjected to:

- Bogus Positive Rate
- Genuine Positive Rate
- Attack Disclosure Time

The model they proposed got a mean of 0.007 with error of 0.01 and scored 100 percent result in TPR. It recognized the attacks in the time of 174–212 ms including the attacks of BASHLITE and Mirai. Three intelligent models with specialization with computer have been developed from the proposed model and relationship b/w them will clarify the doubts on the accuracy of the model. It is doubted that the ability of the model to isolate the attacks from huge traffic maybe from pressure limits, self-instructed, and unquestionable level component extraction limit. To acknowledge the attacks made on the social IoT, Stochastic Slope Plummet learning approach is proposed [9]. This model overpowered other shallow models in every perspective of evaluation when it is subjected to:

- Audit
- Bogus Alert Rate
- Exactness
- F1-measure
- Precision
- Revelation Rate

The model is trained under only NSL-KDD dataset and only subjected to U2R, R2L, and DoS attacks, and under these conditions it outperformed standard simulated intelligent models. To monitor interference area in IoT networks, a better learning model is proposed by the makers which uses the Bidirectional LSTM Repetitive Neural Organization [10]. Only single dataset is used to train this model, it notched 95.7% precision, and it is evaluated under seven estimations:

- Audit
- Distinguishing proof time

- Exactness
- FAR
- F1-score
- Incorrect end-rate
- Precision

The models that where identified are not on par with the proposed method. The makers of the framework came with another model trained using LSTM to identify malware in IoT which targets Opcodes [11]. Previous dataset is used to train the model which got an accuracy of 98% when undergone through: TP, TN, FP, FN, and precision test. It identified malwares counted to 180 and threats counted to 271. The makers again proposed another IoT framework subjected to Programming Characterized Systems administration [12]. For metropolitan networks where security can't be compromised, this model is proposed. To perceive peculiarities, an Intrusion Detection System (IDS) is passed utilizing the RBM. Using artificial intelligence computations and undergoing eight tests—audit, Bogus Negative Rate, Bogus Revelation Rate, exactness, FN, FP, TN, and TP—the proposed model is evaluated. KDD99(dataset containing 1999 attacks which comprises only DoS, R2L, Surveillance, Test and U2R types) is used in training the model, and it goes on to make precision speed of 94% and above. Training the model with the current attacks would make the precision to go much higher. It is identified that IoT applications are more vulnerable and has security threats [13]. To overcome cyberattacks, a model is proposed, and it is tested on: audit, acknowledgment time, exactness, and precision. This model achieved above and below 95% accuracy on the datasets NSL-KDD and UNSWNB-15 respectively. UNSW-NB15 is a dataset with recent attacks while NSL-KDD is a dataset got from making some changes in KDD99. It's observed that the model performs well in prepared dataset than progressing dataset. Besides, the makers in assessment [14] have proposed and completed four significant learning figurings and considered everything against regular simulated intelligence estimations. Further, they have recognized that the hybrid of CNN and LSTM computation have beaten any leftover estimations stood out from significant learning and simulated intelligence counts, with a bewildering precision of 97.16%. Almost, all significant learning models have outmanoeuvred the computer-based intelligence models. Curiously, the dataset was controlled to change the data as it includes significantly unbalanced data. Likewise, limited model appraisal estimations were used, for instance, precision, exactness, and audit. Further, evaluation estimations, for instance, MCC, f-measure, and TPR, may have improved the model. Their framework generally bases on online revelation of association attacks against IoT entries. Using DRNN (Dense Random Neural Network) the framework makers proposed another model to increase the probability of counterattacking future attacks [15]. It mainly focuses on DoS attacks carried on IoT contraption. Results were identical with what we got for previous models. But attacks were limited to absence of rest attack, broadcast attack, deluge attack, TCP SYN, and UDP flood. A similar repeated test was carried out on the model and the result didn't deviate that much. Various organizations are affected by a malware called Ransomware. To detect Ransomware, a model was proposed which accompanies CNN and LSTM. F-measure, FPR, MCC, and TPR were the evaluation tests conducted on the model, and it secured f-extent of 99.6%, and it identified 97.2% of Ransomware. It separated Ransomware in an exact manner how we want it. Duplicated datasets were used while analyzing further. And it works as usual in identifying other attacks like DoS.

9.4 Deep Learning and Big Data Technologies for IoT Security

Here, we are going to identify the relationship b/w our three known zones of evaluation. Significant learning, huge datasets, IoT security, and evaluation strategies were discussed previously and also the methods which maintain them all. We then went on to identify the relationship b/w those. It has showed that either any of those contributed much in past assessments. Immaterial tests have been removed. These points give way for future research. Among three sections on every section, only two assessments we did. The pros and cons are discussed underneath. With improvement in technologies, newly created attacks were carried out on these contraptions, which paved a way for reducing the attacks which resulted in huge downfall. Along these lines, makers of [17] have arranged a significant data structure for interference area using portrayal strategies, for instance, decision trees, conditional Bayes, DNN, random forest, and SVM. The estimations used for appraisal are exactness, audit, sham rate, unequivocally, and assumption time. Apache Flash has been used as a phase for executing interference disclosure in sharp structures using tremendous data assessment. They have declared that the DNN estimation gets the most critical exactness for the unrefined dataset. In light of everything, the most vital precision procured was by the DNN model, yet the exactness is under 80%. Besides, the DNN estimate time is higher and it stood out from various models. What's more, the makers in this examination [18] have inspected the types of progress in hardware, programming, and association geologies, including the IoT, present security perils that require bleeding edge approaches to manage be executed. IDS supported DNN has been proposed in mean time. MLP supported DNN is closed by FFNN. The framework contributed enormously in data development, Apache Flash bundle enlisting stage, and they have been discussed. The Apache Flash pack figuring is set up over the Apache Hadoop One more Asset Mediator YARN. Accuracy, audit, f-score, FPR, precision, and TPR were tests carried out on the model. On both NIDS and HIDS, intelligent models made by man were defeated by this model. Regardless, in the multi-class calcification, the exactness plunges under 90% for explicit attacks in a bit of the datasets. IDS datasets weren't used in the benchmark for DNN.

9.5 Cloud Framework for Profound Learning, Enormous Information Advances, and IoT Security

Application of cloud systems in huge development on data, security of IoT contraptions, and significant learning are dealt under this. Trustable results have been shown by expressive learning in various territories, in any case, significant adjusting maybe extremely computational expansive for gigantic extension applications. This accordingly, powers the joining of extra computational resources as we apply expressive learning on vast application, the resources we had previously would restrict us. Accordingly, cloud structure can be harnessed to settle this test as they hold colossal proportions of resources, for instance, memory, multi-focus GPU and CPU, and move speed. In addition, some cloud systems significantly offer assistance for immense data headways and IoT [19].

9.5.1 Related Works

To overcome the shortcomings in security of IoT framework, quite number of scientists came up with some techniques which help in improving the security and also will help in the future. Most of the frameworks didn't consider ML/DL techniques to remove the loopholes in the security. Recent survey showed IoT frameworks had troubles in the area of access control, application security, confirmation, encryption, and in network security too [20]. Granjal, Silva, and Monteiro [21] underscored the IoT correspondence security after checking on issues and answers for the security of IoT correspondence frameworks. Zarpelão et al. [22] directed a review on interruption discovery for IoT frameworks. Weber [23] zeroed in on lawful issues and administrative ways to deal with the decision if IoT structures fulfill the protection and security prerequisites. Roman, Zhou, and Lopez [24] talked about security and protection in the appropriated IoT setting. These scientists additionally specified a few difficulties that should be tended to and the upsides of the conveyed IoT approach regarding security and protection concerns. Overview [25] explored developing weaknesses and dangers in IoT frameworks, for example, Ransomware assaults and security concerns. Xiao et al. [26] momentarily thought ML techniques are for ensuring information protection and security in the IoT setting. Their investigation likewise showed three difficulties in future ways of ML usage in IoT frameworks (e.g., calculation and correspondence overhead, reinforcement security arrangements, and fractional state perception). Other overview chapters, for example, [27] zeroed in on the employments of information mining and AI techniques for network protection to help interruption recognition. The studies mostly examined the security of the digital area utilizing information mining and AI strategies and predominantly looked into abuse and abnormality recognitions in the Internet [28]. Notwithstanding, rather than different studies, our study presents an extensive audit of bleeding-edge machine and late advances in profound taking in techniques from the point of view of IoT security. This review distinguishes and looks at the chances, benefits, and weaknesses of different ML/DL techniques for IoT security. We talk about a few difficulties and future headings and present the distinguished difficulties and future bearings based on looking into the potential ML/DL applications in the IoT security setting, subsequently giving a helpful manual to scientists to change the IoT framework security from just empowering a safe correspondence among IoT parts to start to finish IoT security-based wise methodologies. In [29], the author proposed and assessed a novel recognition technique that separates social previews from the organization and auto-encoders have utilized to detect unusual traffic in the network from traded-off gadgets. The significant downside of utilizing unaided AI calculations for recognition issues is traffic in the network, the greater part of the streams are ordinary and abnormalities like assaults and exceptions are uncommon, which contrarily influences achievement rates and the location of peculiarities. Thus, better outcomes are normal with administered methods. Then again, many directed learning calculations are utilized to distinguish assaults and are prepared on datasets with names showing whether the examples have been pre-delegated assaults or not. With the help of ML techniques and ANN and SVM, non-T or traffic attacks are identified in the UNBCIC datasets [30] and also from whitelist it recognized IoT contraption types. In order to segregate the information got on traffic in the network, Random Forest technique is used [31]. Another work that has a similar method to manage our assessment was presented by [32] in the principal chapter which proposed the Bot-IoT dataset. IoT dataset was assessed by LSTM, RNN, and SVM models, anyway in their assessment they didn't choose the adversarial force of their models. In our work, while we use a comparable Bot-IoT dataset presented in [33], we base on removing new features from the dataset and evaluating unmistakable man-made intelligence counts on this dataset [34] is another assessment that used the BoT-IoT dataset.

9.6 Motivation of the Proposed Methodology

- To improve the security in the sensational data and to overcome from various attacks using fast Fourier transform with Multi-layered Deep Neural Network (MDNN).
- To resort (protecting the sensational data from the various attacks) the data from cloud infrastructure are of large resource to secure the data from various attacks using IDS with deep neural network.
- To improve classified accuracy using Ensemble classification techniques such as and along with the relevant images retrieval using K-means classification technique with appropriate similarity among the query image.

9.7 Research Methodology

The suggested Deep Learning Protection framework in Figure 9.1 is a distributed, lightweight, agent-based intrusion detection system. The model architecture is similar to that proposed in, but the trust management and fault tolerance mechanisms are completely different. The agents are seen as autonomous, reflexive, constructive, and cooperative actors in the proposed approach. They are in charge of the study's data collection, interpretation, and inference. The agents use an inference approach that uses the collected data as evidence in an MDNN network. For precision and fault tolerance, such as dealing with the possibility of certain agents being compromised, monitoring, and analysis research is duplicated.

Recent IoT technologies and advanced deep learning techniques used in implementing framework are described in Figure 9.2. Development in analysis got interest, as when various sectors started to create vast amount

9.7 Research Methodology

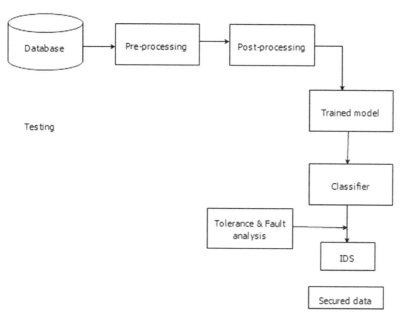

Figure 9.1 Overall architecture of the fault tolerance deep analysis secured data.

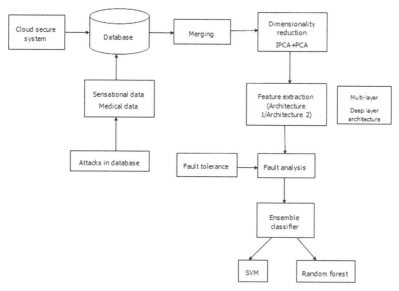

Figure 9.2 Implementation architecture of the sensational attacks analyzing fault tolerance ensemble classification using MDNN.

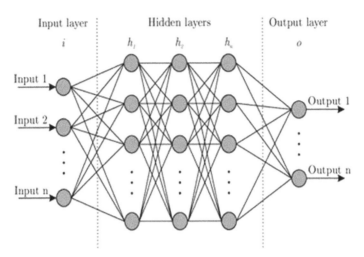

Figure 9.3 MDNN architecture.

of data. In Figure 9.3 discusses about MDNN architecture and this framework has 3 stages: Using the dataset taken from the cloud as input is the first stage; analyzing the fault in the classification, obtained using deep learning architecture; on the basis of accuracy, precision, recall, metrics, and F1 score, the framework will be evaluated. In a short period of time, with significant number of higher accuracy, they claimed their model will analyze a large amount of data with less computational power. It is said that at every stage, the model gives 95% of accuracy, F1 score, recall, and precision. The framework proposed has been embodied with insufficient IoT data.

9.7.1 Dimensionality Reduction

There are different methods that can be used in the pre-processing technique for DR. For the following advantages it offers, the DR is considered:

- Reducing the dimensions decreases required memory to save the data.
- Reduced training or computing times require a smaller number of dimensions.
- If data has several dimensions, most of the algorithms in the process of feature extraction don't really handle well.
- The multicollinearity between various data features is well dealt with by DR techniques and redundancy between the features is eliminated.
- It aids in imagining data's lower dimensions at last.

9.7.2 Independent Component Analysis

A relatively new statistical and computational methodology for data processing is independent component analysis (ICA). ICA emerged from the community of signal processing, where it was developed as a powerful technique for separating blind sources. It is possible to write the basic ICA model for feature transformation as:

$$s_t = u x_t$$

where,
$x_t \rightarrow$ observed feature vectors (matrix with dimensions $n \times p$)
$s_t \rightarrow$ new independent vectors (matrix with dimensions $n \times p$)
u$\rightarrow$ de-mixing matrix (matrix with dimensions $n \times p$)

The most non-Gaussian direction is the first IC. A new statistically independent "non-Gaussian" directions coordinate system is found by x_t. Initially, the most non-Gaussian direction is identified by the algorithm through an iterative process. This process continues finding 2nd, 3rd, ..., etc., most non-Gaussian directions using the previous direction which is independent. Independent vectors (matrix with dimensions $n \times p$) are determined from data vectors (matrix with dimensions $n \times p$) so that original data can be projected into u (independent components) estimated from the data. s_t's join probability density function equals implied by statistical independence. The number of s_t and the dimensions of x_t will be reduced consequently using PCA. Here, both PCA and the selected PCs have the same numbers. Upon completing the data whitening process, independent components and transformation matrix are estimated using fixed point-algorithm. Mutual information is called a measure of the dependency between random variables. Maximizing their negentropy is equal to minimizing the mutual information between the components. Negentropy can be expressed approximately as follows in the fast ICA:

$$J_G(s_{t(i)}) \approx [E\{G(s_{t(i)})\} - E\{G(V)\}]^2$$

where,
G $\rightarrow$ non-quadratic function,
V $\rightarrow$ Zero mean and unit variance Gaussian variable,
$s_{t(i)} \rightarrow$ n-dimensional vector, one of the rows taken from the matrix u.

Below optimization problem is resulted from substituting in $s_{t(i)} = u_i^T x_t$ Eq.

$$\text{Maximize } \Sigma_i^t s_{t(i)} J_G(\mu_i) = [E\{G(\mu_i^T x)\} - E\{G(V)\}]^2$$
$$\text{Subject to } E\{(\mu_i^T x)^2\} = 1 \; i = 1, 2, \ldots n$$

We can get one new independent component from the solution to this optimization problem, solved with the help of Fast ICA algorithm. s_t (i.e., reduced IC matrix) can be calculated based on this.

9.7.3 Principal Component Analysis

In Principal Component Analysis, new features (aka Principal Components) are added by converting the existing features in a dataset. The new features (aka principal components) are combinations of initial variables linearly and are categorized in a way that the highest variance is represented by the first principal component of the data. The variance remained are represented by subsequent new features, which doesn't supplement the first principal component. Redundant features are eliminated by SVD (i.e., singular value decomposition) through components decomposed from the original matrix, based on Eigenvalues and Eigenvectors. The above transformation is made keeping in mind that the greatest possible variance should be the first principal component.

$$x_t \in R^n \text{ and } \sum_{t=1}^{p} Rx_t = 0 | \; t = 1, 2 \ldots p$$

where,

$x \to$ given input matrix with dimension $n \times p$ and $n < p$

Above is the mathematical idea behind PCA.

Obtaining the Eigenvalues in the given problem is the way by which this vector space is transformed into new vectors by PCA as follows:

$$\lambda_i \mu_i = c \mu_i \; i = 1, 2, \ldots, n$$

where,
$\lambda_i \geq 0 \to$ one among n Eigenvalues of $n \times p$ covariance matrix(c),

$$c = \left(\frac{1}{p} \left(\sum_{t=1}^{p} x_t x_t^T \right) \right)$$

$\mu_i \to$ Eigenvector of λ_i

$$y_t(i) = \mu_{1i}x_1 + \mu_{2i}x_2 + \ldots\ldots + \mu_{ni}x_n = \mu_i^T x_t \quad i = 1, 2, \ldots, n$$

which is an orthogonal transformation of $y_t(i)$.

Among all $y_t(i)$ in the data set, the vector which has maximum variance is chosen by the highest Eigenvalue. With the help of first many Eigenvectors got from sorting them in non-increasing order w.r.t., Eigenvalues, new vectors of PCs got reduced from initial vector space of PCs I k < n.

9.7.4 Cloud Architecture

There are only a few blocks, that is, cloud servers (C1, C2) and clients (multiple). The authenticated client encrypts the trained images and stock unreadable image data on cloud servers; a content-based query image will be taken from the users by the server. To store the unreadable encrypted sub images, two cloud servers, C1 and C2, as shown in Figure 9.4 are used, respectively. The pre-trained output images of the CRNN model typically undergo some heterogeneous operations on the linear layers. To evaluate the computations with high protection on the non-linear layers, C1 retains its private key sk and IIES operations are performed in C2. Where the feature extracted from the model is shared by C1, C2 cloud servers and images are encrypted. The encrypted images uploaded by the trained model users are converted by a user to the encrypted data format and a test image is identified. However, the trained image I user gets the encrypted image I_a by summating a stochastic matrix I_b along with it. However, I_a is transferred to the cloud server C1, and I_b is transmitted to C2. Exploratory SNN images are created in a similar pattern by a client over the test image trapdoor. Subsequently, accepted I_a, C1 carries out Improved Image Encryption System (IIES) protocols with C2 to convert the unintelligible image, I_b to the primary image with key sk.

9.7.5 Encryption Decryption Using OTP

The concept of: For a message, an OTP (One-time pad) can be used as a key only once; is an encryption method used in cryptography and information theory. In this method, equivalent to the length of the message to be encrypted, a random key is generated without repetition. Bitwise XOR is performed b/w plaintext and key and generated output is the ciphertext. C = P⊕K, where P = Plaintext, ⊕ = Exclusive-Or, K = Key, and C = Ciphertext is the

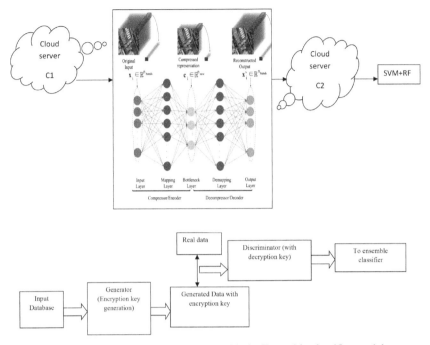

Figure 9.4 The system architecture with the Ensemble classifier model.

expression used in OTP, where P → Plaintext, ⊕ → XOR, K → Generated Random Key, and C → Ciphertext. Performing the same operation b/w C and K gives us back P, this is called decryption. Either OTP or random key used is found, the system is vulnerable.

The Mathematical Proof of OTP Security

Let,
  L → # bits in the plaintext string and

$$i \in (1, L)$$

Then,
  p_i → i^t bit in the plaintext string,

  c_i → i^t bit in the ciphertext string,

  k_i → i^t bit in the key string,

  $P(p_i)$ → probability that p_i was sent,

9.7 Research Methodology

$P(p_i|c_i) \rightarrow$ probability that p_i was sent given that c_i was observed.

If $P(p_i|c_i) = (p_i)$, then the system will be perfectly secret. This section deals about why completely secret, the OTP system is. Performing XOR b/w plaintext data bits and key bits, is the standardized way.

Bitwise XOR-ing key and plain text give us ciphertext from sender side:

$$c_i = p_i \oplus k_i \rightarrow P(p_i) \; [1]^i$$

Due to the unpredictability and random nature of OTP, we make the following assumptions:

1. Probability of detecting a key bit is same for each key in OTP and in One Time Key.
2. We knew next to nothing from the previous bits of the key in determining the next bit.

Distribution of P(Ci):

P(ci = 1) = P(ci = 1 | ki = 1)

P(ki = 1) + P(ci = 1 | ki = 0)

P(ki = 0) by the definition of condiional probability = P(pi = 0)

P(ki = 1) + P(pi = 1)

P(ki = 0) by equation (1)

= P(pi = 0) (1/2) + P(pi = 1) (1/2) by equation (2)

= (1/2) [P(pi = 0) + P(pi = 1)]

regrouping = 1/2

since pi can only be 1 or 0

P(ci = 0) = P(ci = 0 | ki = 1) P(ki = 1) + P(ci = 0 | ki = 0)

P(ki = 0) by the definition of conditional probability

= P(pi = 1) P(ki = 1) + P(pi = 0) P(ki = 0) by equation (1)

= P(pi = 1) (1/2) + P(pi = 0) (1/2) by equation (2)

= (1/2) [P(pi = 1) + P(pi = 0)]

regrouping = 1/2 since pi can only be 1 or 0

Distribution of P(ci | pi):

If pi = 0: P(ci = 0 | pi = 0) = P(ki = 0) by equation (1)

= 1/2 by equation (2)

P(ci = 1 | pi = 0) = P(ki = 1) by equation (1)

= 1/2 by equation (2)

If $p_i = 1$: $P(c_i = 0 \mid p_i = 1)$

= $P(k_i = 1)$ by equation (1)

= 1/2 by equation (2)

$P(c_i = 1 \mid p_i = 1)$

= $P(k_i = 0)$ by equation (1)

= 1/2 by equation (2)

So the above derivation implies $P(c_i \mid p_i) = P(c_i)$. $P(p_i) = P(p_i \mid c_i)$ implies a system is perfectly secret. With the use of joint and conditional probability and with $P(c_i)$ and $P(p_i \mid c_i)$ we can write $P(c_i \mid p_i) P(p_i)$ in the following ways:

$P(p_i \text{ and } c_i) = P(c_i \mid p_i) P(p_i)$ and $P(p_i \text{ and } c_i) = P(p_i \mid c_i) P(c_i)$

$P(p_i \mid c_i) P(c_i) = P(c_i \mid p_i) P(p_i)$

Combining the above equations gives,
$\quad P(c_i \mid p_i) = P(c_i)$

Cancelling the terms using the above proof leaves us with:
$\quad P(p_i \mid c_i) = P(p_i)$ which implies perfect secrecy.

(i) Encryption Phase—To deny access to unauthorized, data is converted to cipher in this process.

(ii) Decryption Phase—Decrypting the encrypted text is done in this process. This word may be used to specify a method of manually decrypting the data or using the correct codes or keys to decrypt the data. The random number generated from two sources is retrieved. Creating a random number from a single source, maybe gets predictable. Therefore, by having the random numbers from two sources, we implemented this scheme. All computer generated random number are pseudo-random, meaning a mathematical formula is used to generate them predictably. For certain purposes, this may be good, but it does not provide full protection. Our first source of randomness is ambient noise, which is better than the commonly used pseudorandom number algorithms in computer programs for many purposes. Another factor has been incorporated so that by throwing a ten-sided dice, we can get full randomness.

9.7.6 OTP Algorithm

The Encryption Algorithm

Step 1: Load file
Step 2: Read the content of the file and store in an array list Plain
Step 3: Calculate the number of blocks in the file and the offset (Plain / 64 = blocks; Plain % 64 = offset)
Step 4: Generate 512 bit random key
Step 5: Generate a random integer for each file block and store in an array
Step 6: Shift the key four times by four of the above generated random numbers and store the shifted keys as s1, s2, s3, and s4
Step 7: Generate a one-time pad (OTP) with the key and the random integers

a: Get the total size of the plain text
b: Shift the key with a random integer
c: Inverse the shifted key NOT (shift key)
d: Repeat steps b and c for each block, append it to the one-time pad array
e: For the offset append one (1) to the pad and append zeros till the pad is complete
f: Return one-time pad = OTP Step

step 8: Break the plain text into blocks of 512 bit each and perform a Cipher Block Chaining (CBC) with each of the previously shifted keys

a: Exclusive Or the block and key, block XOR key = text1
b: Exclusive Or text1 and s1, text1 XOR s1 = text2
c: Exclusive Or text2 and s2, text2 XOR s2 = text3
d: Exclusive Or text3 and s3, text3 XOR s3 = text4
e: Exclusive Or text4 and s4, text4 XOR s4 = text5
f: Append text5 to the array text
g: Repeat steps a through f for each block

Step 9: Exclusive Or each bit in OTP and the bit in text, OTP XOR text = cipher
Step 10: Convert the binary values in cipher to character and store in file
Step 11: Store the random key into the key file and append it with all random numbers used for shifting
Step 12: End B.

The Decryption Algorithm

Step 1: Load file
Step 2: Read the content of the file and store in an arraylist

Step 3: Calculate the number of blocks in the file and the offset (Plain / 64 = blocks; Plain % 64 = offset)
Step 4: Read the content of the key file, store the key into an array
Step 5: Store the shift integers into another array
Step 6: Shift the key four times by 4 of the shift integers and store as s1, s2, s3, and s4
Step 7: Generate a one-time pad with the key and the random integers

a: Calculate the total size of the plain text
b: Shift the key with the shift integers gotten from the key file
c: Inverse the shifted key NOT (shift key)
d: Repeat steps b and c for each block, append it to the one-time pad array
e: For the offset append one (1) to the pad and append zeroes till the pad is complete
f: Return one-time pad = OTP

Step 8: Convert the cipher text to binary values = text
Step 9: Exclusive Or each bit in OTP and the bit in text, OTP XOR text = text1
Step 10: Break the cipher text into blocks of 512 bit each and perform a Cipher Block Chaining (CBC) with each of the previously shifted keys

a: Exclusive Or the tex1 and s4, text1 XOR s4 = text2
b: Exclusive Or text2 and s3, text2 XOR s3 = text3
c: Exclusive Or text3 and s2, text3 XOR s2 = text4
d: Exclusive Or text4 and s1, text4 XOR s1 = text5
e: Exclusive Or text5 and key, text5 XOR key = text6
f: Append text6 to the array plain
g: Repeat steps a through f for each block

Step 11: Convert the binary values to character and store into the plain text file
Step 12: End

9.7.7 Ensemble Classifier SVM, Random Forest Classification

SVM:

SVM (Support Vector Machine) is a supervised learning algorithm. The problems involving classification/regression can be solved by this but predominantly in classification. Each data item is represented by an n-dimensional space point in the SVM algorithm where n → # features we have. Each feature is pointed by a unique coordinate. Classification is done by finding a hyper-plane which differentiates the two classes for a good extent.

Support Vectors are simply individual observation coordinates. The SVM classifier is a boundary that divides the two classes most effectively (hyperplane/line). SVMs are a set of similar techniques of supervised learning used in classification and regression. They belong to a generalized linear classification family. The empirical classification error is simultaneously minimizing by a special SVM property and the geometric margin is maximizing. Structural Risk Minimization is the place where SVM concentrates a lot. SVM maps the input vector to a higher dimensional space where a maximum hyperplane separation is created. To separate the data, two Parallel hyperplanes are created on each side of the main hyperplane. Parallel hyperplanes are divided by a hyperplane, which maximizes the distance b/w the two. Assumption: Generalized error and distance b/w Parallel hyperplanes are in indirect relationship. Consider the following data points:

$$\{(x_1, y_1), (x_2, y_2), (x_3, y_3), (x_4, y_4), ...(x_n, y_n)\} \qquad (9)$$

where,
x_n → p-dimensional real vector,
n → # sample,
$y_n = 1/-1$ → Has value 1 or −1.

To safeguard against variable (attributes) with greater variance, the scaling is necessary. Dividing hyperplane has the equation,

$$w. x + b = o \rightarrow [1]$$

where,
b → scalar,
w → p-dimensional vector.

w and hyperplane dividing it are ⊥. We may increase the margin by adding the offset parameter b. The hyperplane is forced to pass through the origin in the absence of b, limiting the solution. We are interested in SVM and parallel hyperplanes, as we are interested in the maximum margin. Parallel hyperplanes can be described by equation w.x + b = 1, w.x + b = −1. We can pick these hyperplanes so that there are no points between them if the training data is linearly separable, and then try to optimize their distance. $2/|w|$ is the distance b/w the hyperplanes by geometry interpretation. So the next thing to do, is minimizing |w| is ensured for excitation of data points.

$$|y_i(w \cdot x_i - b)| \geq 1, 1 \leq i \leq n \rightarrow [2]$$

SVs (aka Support Vectors) are the samples along the hyperplanes. M = 2 / |w| is the largest margin, separating hyperplane which has this margin, specifies training data points closest to the support vector machine.

A canonical hyperplane with maximum margin is given by:

$$y_j [w^T \cdot x_j + b] = 1, j = 1 \rightarrow (3)$$

The following are the constraints that OCH should follow every data:

$$y_i[w^T \cdot x_i + b] \geq 1; i = 1,2 \ldots 1 \rightarrow (4)$$

where,

l → # training data point

Minimizing $||w||^2$ w.r.t. inequality constraints will help us in finding optimal separating hyperplane which has maximal margin:

$$y_i [w^T \cdot x_i + b] \geq 1; i = 1, 2 \ldots \ldots l$$

Lagrange l function's saddle points will solve this optimization problem.

$$L_P = L(w, b, \alpha) = 1/2 ||w||2 - \sum \alpha_i (y_i (w^T x_i + b) - 1) \; i = 1 \; l = 1/2 \; w^T w - \sum \alpha_i (y_i(w^T x_i + b) - 1) \rightarrow (5)$$

where,

$\alpha_0 \rightarrow$ Lagrange's multiplier

Minimizing Lagrange w.r.t. w and b and maximizing w.r.t. $\alpha_0 (\geq 0)$ makes it necessary to find the optimal saddle points (w0, b0, α0).

Any of the following will solve the problem:

1. Primal form (which has the terms w & b)
2. Dual form (which has the term α_0)

αi ($\alpha i \geq 0$) (4) and (5) are convex, and they hold KKT conditions, which is sufficient to maximize the $\alpha_i(4)$. Applying partial differentiation on $\alpha_i(5)$ w.r.t. saddle points (w0, b0, α0).

$$\partial L / \partial w0 = 0 \; l \; i.e \; w0 = \sum \alpha i \; yi \; xi \rightarrow (6)$$
&
$$\partial L / \partial b0 = 0 \; l \; i.e \sum \alpha i \; yi = 0 \rightarrow (7)$$

We get the dual form from the primal form by substituting (6) and (7) in (5).

$$Ld(\alpha) = \sum \alpha_i - 1/2 \sum \alpha_i \alpha_j y_i y_j x_i^T x_j = 1 \rightarrow (8)$$

Maximizing Ld (dual Lagrangian) w.r.t. non-negative a_i and equality constraints will help us in identifying the optimal hyperplane as follows:

$$\alpha_i \geq 0, i = 1, 2, \ldots 11 \sum \alpha_i y_i = 0 \; i = 1$$

Ld(α) (dual Lagrangian) depends on the input pattern's scalar products ($x_i^T x_j$) only and is expressed in terms of training data.

9.7.8 Random Forest

A classifier which reach single output through combination of multiple decision trees output is called Random Forest.

$$\{h(x, \in k) k = 1, 2, \ldots\}$$

where,
$\{\Theta k\} \rightarrow$ identically distributed independent random vectors.

At input x, the most favored class is cast a unit vote by each tree. Breiman followed below steps, to produce a single tree in Random forest:

- From the earliest data, at random sample N cases with replacement, where N is # cases in the training set. The growing tree will have new cases as training set.
- At each node out of the M, at random m variables are selected (where, M $\rightarrow$ input variables and m $<<$ M). The node is separated at which these m variables are separated at best. During growth of the forest, m is kept as constant.
- Without pruning, each tree attains its full growth.

In the forest, collective trees are brought into through this method; The Ntree parameter decides # trees in advance. The variable m is also named as mtry or k literally. # instances in the leaf node (aka parameter node size) controls the tree's depth and its default value is 1. Aiming to classify a new instance, the above steps are followed to train/built the forest and once it's done, each grown tree in the forest, undergoes the same run. Votes are recorded for every classification done by each tree on the new instance. The new instance is classified as the class, which gets the maximum votes taking account of all trees.

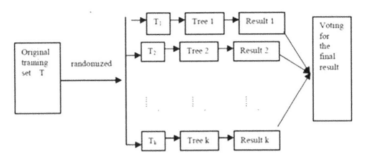

Figure 9.5 Random forest classification.

For a given input variable, each tree has right to vote to select the best classification result. Specific process is shown in Figure 9.5.

We refer Forest RI generated decision tress as Random Forest, here on. Initial instances of about one-third are left out for each tree, when an active sample set is taken out by sampling with replacement. These instances are termed as Out-of-Bag (OOB) data. In the forest, each tree's error is estimated from its own OOB data set and named as OOB error estimation. In Random Forest, the following are:

- PE * = P x,y (mg(X,Y)) < 0 → Generalized error
- mg (X,Y) = avk I(hk (X) = Y) − max j≠Y avk I(hk (X) = j)→ margin function

The measure of the amount by which avg # votes at (X, Y) for the right class exceeds the avg vote for any other class is the margin function.

- S = E X, Y (mg (X, Y)) → Strength of Random Forest
- PE* ≤ ρ (1 − s2) / s2 → Upper bound of Generalized error where, $\underline{\rho}$ → mean value of correlation b/w base trees.

Thus, it implies, accuracy of Random Forest is in direct relationship with diversity and accuracy of the base decision trees.

9.8 Performance Metrics

To evaluate performance of the system, a pack of statistical metrics will be followed in this section below.

The performance analysis of the submitted method is illustrated below. Parameters to be considered for evaluation are recall, AUC, precision, accuracy, and F1 score. Various performance metrics have been calculated using the classified output.

Accuracy: It shows the percentage of correctly classified instances during classification.

$$Accuracy\, rate = \frac{TruePositive + TrueNegative}{TotalInstances} * 100$$

Precision: It measures what proportion of data transmitted to the network had an intrusion. The predicted positives (Network predicted as the intrusion is TP and FP) and the network having an intrusion are TP. This is used to measure the quality and exactness of the classifier as shown below:

$$Precision = \frac{TruePositive}{TruePositive + FalsePositive}$$

Recall: The recall is the ratio Real Positives which are correct Predicted Positive and is defined as

$$Recall = \frac{TruePositive}{TruePositive + FalseNegative}$$

F1 Score: The F1 Score reflects the average accuracy and recall value. Definition of precision is the judgment of accuracy whereas recall is detecting the sample instance based on the attribute called faulty or non-faulty.

$$F1 - Measure = \frac{2 \times Precision \times Recall}{Precision + Recall}$$

Output of model-generated belongs to one of the following four categories which are used to calculate the previous metrics:
- False Negative (FN) → label—attack; prediction—normal.
- False Positive (FP) → label—normal; prediction—attack.
- True Negative (TN) → both label and prediction are normal.
- True Positive (TP) → both label and prediction are attack.

9.9 Dataset Description

Cyber security datasets used in network intrusion detection to benchmark the performance assessment and comparison of IDS, include regular and unusual traffic in the network in each dataset. Owing to the privacy concerns of organizations, infrastructure data on traffic in the real-world network is largely internal as well as inaccessible. Publicly accessible datasets, however, enable

Table 9.1 Comparison of the existing and the proposed Ensemble classifier using NSL-KDD dataset

S.NO	Metric	DBN (%)	SVM (%)	Random Forest (%)	Proposed Ensemble (%)
1	Area Under ROC	96.55	94.95	97.25	98.56
2	Precision	96.80	92.62	96.87	97.27
3	Recall	88.47	86.22	91.98	92.17
4	F-Measure	92.45	89.30	94.36	94.65
5	Accuracy	92.11	93.24	93.45	95.12

the researcher to obtain a true assessment of IDS results. These datasets lack variety in traffic, large volumes of traffic and new types of assaults such as low footprint. Therefore, in order to give logical and fair competition to other modern detecting approaches, old, and new public datasets are used.

Tavallaee et al. proposed newly renovated KDD Cup 99 dataset called NSL-KDD [39]. Amount of data about traffic in the network, is around 4 GB in this dataset. Network flow's different features are deployed by 41 attributes which describes each raw and to indicate the type of flow, each is assigned with a label such as normal or any specific attack. As it is used in testing unusual detection models, we selected this dataset, and also it has numerous attributes and instances.

Table 9.1 shows some of the observation of the KDD Cup 99 Dataset, the outcome of the classifier has been estimated from the instances of the Real world IoT data, then classifying the instances with the same observation, then the performance measures of various techniques of DBN, SVM, and Random Forest are compared with Proposed_Ensemble. The table displays accuracy, area under ROC, F-Measure, precision, and recall performance comparisons. It has been analysed from the Actual and predicted value which are taken from the objective of classes of various attacks in IDS.

In terms of accuracy, Figure 9.6 indicates the contrast of different approaches. It is an accuracy comparison for KDD Cup 99 Dataset between existing and proposed techniques. As shown in Figure 9.6, the Proposed Ensemble achieves accuracy with a maximum percentage than the existing techniques. Whereas DBN, SVM, and Random Forest approaches have resulted in the worst performance by furnishing a minimum of training accuracy values of about 92.11%, 93.24%, and 93.45%. Finally, the Proposed_Ensemble technique operates more efficiently when compared with other models by acquiring the maximum accuracy value of training values of about 95.12%.

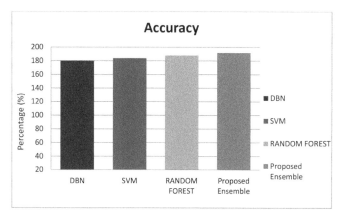

Figure 9.6 Accuracy comparison of various existing techniques with Proposed Ensemble by applying KDD Cup 99 Dataset

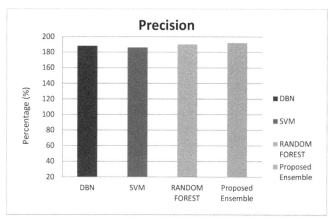

Figure 9.7 Precision comparison of various existing techniques with Proposed Ensemble by applying KDD Cup 99 Dataset.

In terms of precision, Figure 9.7 illustrates the comparison of various methods. It is a precision comparison between current and proposed techniques for the KDD Cup 99 Dataset. As shown in Figure 9.7 above, with a maximum percentage of current techniques, the Proposed Ensemble achieves precision. Whereas by having a minimum of training precision values of about 96.80%, 92.62%, and 96.87%, the DBN, SVM, and Random Forest methods have resulted in the worst results. Finally, by acquiring the full precision value of training values of about 97.27%, the Proposed_Ensemble system performs more effectively compared to other models.

144 Fault Tolerance-Based Attack Detection

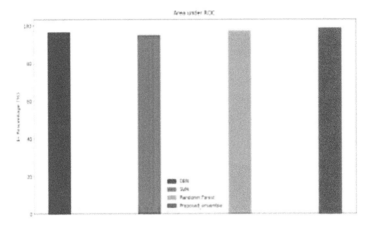

Figure 9.8 Area under ROC comparison of various existing techniques with Proposed Ensemble by applying KDD Cup 99 Dataset

Figure 9.8 represents the AU-ROC. This illustrates the comparison of various methods. It is an AU-ROC comparison between current and proposed techniques for the MIB Dataset. As shown in Figure 9.8 above, with a maximum percentage of current techniques, the Proposed Ensemble achieves AU-ROC. Whereas by having minimum of training AU-ROC values of about 96.55%, 94.95%, and 97.25%, the DBN, SVM, and Random Forest method have resulted in the worst results. Finally, by acquiring the full F1-score value of training 98.56%, the Proposed Ensemble system performs more effectively compared to other models.

With respect to recall, Figure 9.9 shows the comparison between different approaches. It is a KDD Cup 99 Dataset recall comparison between current and proposed techniques. As shown in Figure 9.9 above, with a maximum percentage of current techniques, the proposed Ensemble achieves recall. Whereas DBN, SVM approach have resulted in the worst performance by furnishing minimum Recall values of about 88.47%, 86.22% respectively for training. Whereas, Random Forest gradually increased the Recall value to about 91.98% for training. Finally, the Proposed Ensemble technique operates more efficiently when compared with other models by acquiring the maximum Recall value of 92.17% for training.

Figure 9.10 represents the F-Measure. This illustrates the comparison of various methods. It is an F-Measure comparison between current and proposed techniques for the KDD Cup 99 Dataset. As shown in Figure 9.10, with a maximum percentage of current techniques, the Proposed Ensemble

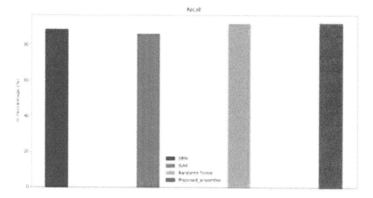

Figure 9.9 Recall comparison of various existing techniques with Proposed Ensemble by applying KDD Cup 99 Dataset.

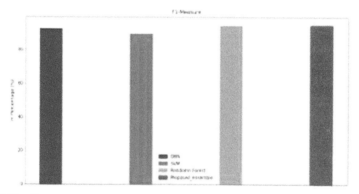

Figure 9.10 F1-Measure comparison of various existing techniques with Proposed Ensemble by applying KDD Cup 99 Dataset

achieves F1-Measure. Whereas by having minimum of training F1-Measure values of about 92.45%, 89.30%, and 94.36%, the DBN, SVM, and Random Forest methods have resulted in the worst results. Finally, by acquiring the full F1-Measure value of the Proposed Ensemble system performs 94.65% more effectively compared to other models.

9.10 Conclusion

The reason for increasing number of security threats for IoT contraptions is their rise in popularity. Botnets such as Mirai and Carna are the common ones which attacked IoT contraptions using vulnerability. Huge volumes of different type data are produced by IoT contraptions. So the research suggests that modern-day solutions are in requirement. So, this chapter proposes the model designed between cloud architectures. The input dataset will be a medical database or sensational dataset; the data has been from IoT. Before this IoT, the cloud structure has been projects with security, and the output of deep learning architecture with ensemble classifiers will have the IDS for security. The threshold limit has been established.

Similarly, those data beyond the threshold limit will be sensational. Both the cloud architecture will have a similar private key, whether the data is securely transmitted or not. Each information will have its threshold limit, and this has been analyzed, and categorized classified output is from deep learning ensemble classifiers. The experimental results show effective data transmission with higher accuracy when compared with existing techniques.

References

[1] Hussain, Fatima, et al. "Machine learning in IoT security: Current solutions and future challenges." *IEEE Communications Surveys & Tutorials* 22.3 (2020): 1686–1721.

[2] Canedo, Janice, and Anthony Skjellum. "Using machine learning to secure IoT systems." *2016 14th annual conference on privacy, security and trust (PST)*. IEEE, 2016.

[3] Mohanta, Bhabendu Kumar, et al. "Survey on IoT security: challenges and solution using machine learning, artificial intelligence and blockchain technology." *Internet of Things* (2020): 100227.

[4] Meidan, Yair, et al. "Detection of unauthorized IoT devices using machine learning techniques." *arXiv preprint arXiv:1709.04647* (2017).

[5] Al-Garadi, Mohammed Ali, et al. "A survey of machine and deep learning methods for internet of things (IoT) security." *IEEE Communications Surveys & Tutorials* 22.3 (2020): 1646–1685.

[6] M. Ahmed, A.N. Mahmood, J. Hu, *A survey of network anomaly detection techniques*, J. Netw. Comput. Appl. 60 (2016) 19–31.

[7] M. Nobakht, V. Sivaraman, R. Boreli, A host-based intrusion detection and mitigation framework for smart home iot using openflow, in: 2016 *11th International Conference on Availability, Reliability and Security (ARES)*, IEEE, 2016, pp. 147–156.

[8] J. Saxe, K. Berlin, Deep neural network based malware detection using two dimensional binary program features, in: 2015 *10th Internationa Conference on Malicious and Unwanted Software (MALWARE)*, IEEE, 2015, pp. 11–20.

[9] I. Kara, M. Aydos, Static and dynamic analysis of third generation cerber ransomware, in: 2018 *International Congress on Big Data, Deep Learning and Fighting Cyber Terrorism (IBIGDELFT)*, IEEE, 2018, pp. 12–17.

[10] J.M. Ceron, K. Steding-Jessen, C. Hoepers, L.Z. Granville, C.B. Margi, *Improving iot botnet investigation using an adaptive network layer*, Sensors 19 (3) (2019) 727.

[11] C. Kolias, G. Kambourakis, A. Stavrou, J. Voas, *Ddos in the iot: Mirai and other botnets*, Computer 50 (7) (2017) 80–84.

[12] S. Vashi, J. Ram, J. Modi, S. Verma, C. Prakash, Internet of things (iot): A vision, architectural elements, and security issues, in: 2017 *International Conference on I-SMAC (IoT in Social, Mobile, Analytics and Cloud) (I-SMAC)*, IEEE, 2017, pp. 492–496.

[13] I. Andrea, C. Chrysostomou, G. Hadjichristofi, Internet of things: Security vulnerabilities and challenges, in: 2015 *IEEE Symposium on Computers and Communication (ISCC)*, IEEE, 2015, pp. 180–187.

[14] S. Son, V. Shmatikov, The hitchhiker's guide to dns cache poisoning, in: *International Conference on Security and Privacy in Communication Systems*, Springer, 2010, pp. 466–483.

[15] P. De Ryck, L. Desmet, F. Piessens, W. Joosen, Secsess: Keeping your session tucked away in your browser, in: *Proceedings of the 30th Annual ACM Symposium on Applied Computing*, ACM, 2015, pp. 2171–2176.

[16] K. Sonar, H. Upadhyay, A survey: *Ddos attack on internet of things*, Int. J. Eng. Res. Dev. 10 (11) (2014) 58–63.

[17] A. Bijalwan, M. Wazid, E.S. Pilli, R.C. Joshi, *Forensics of random-udp flooding attacks*, J. Netw. 10 (5) (2015) 287.

[18] Y.G. Dantas, V. Nigam, I.E. Fonseca, A selective defense for application layer ddos attacks, in: 2014 *IEEE Joint Intelligence and Security Informatics Conference*, IEEE, 2014, pp. 75–82.

[19] J. Krupp, M. Backes, C. Rossow, Identifying the scan and attack infrastructures behind amplification ddos attacks, in: *Proceedings of the 2016 ACM SIGSAC Conference on Computer and Communications Security*, ACM, 2016, pp. 1426–1437.

[20] A. R. Sfar, E. Natalizio, Y. Challal, and Z. Chtourou, "*A roadmap for security challenges in the Internet of Things,*" Digital Communications and Networks, 2017

[21] J. Granjal, E. Monteiro, and J. S. Silva, "Security for the internet of things: a survey of existing protocols and open research issues," *IEEE Communications Surveys & Tutorials*, vol. 17, no. 3, pp. 1294–1312, 2015.

[22] B. B. Zarpelão, R. S. Miani, C. T. Kawakani, and S. C. de Alvarenga, "A survey of intrusion detection in Internet of Things," *Journal of Network and Computer Applications*, vol. 84, pp. 25–37, 2017.

[23] R. H. Weber, "Internet of Things–New security and privacy challenges," *Computer law & security review*, vol. 26, no. 1, pp. 23–30, 2010

[24] R. Roman, J. Zhou, and J. Lopez, "On the features and challenges of security and privacy in distributed internet of things," *Computer Networks*, vol. 57, no. 10, pp. 2266–2279, 2013.

[25] I. Yaqoob et al., "The rise of ransomware and emerging security challenges in the Internet of Things," *Computer Networks*, vol. 129, pp. 444–458, 2017.

[26] L. Xiao, X. Wan, X. Lu, Y. Zhang, and D. Wu, *"IoT Security Techniques Based on Machine Learning,"* arXiv preprint arXiv:1801.06275, 2018.

[27] P. Mishra, V. Varadharajan, U. Tupakula, and E. S. Pilli, "A Detailed Investigation and Analysis of using Machine Learning Techniques for Intrusion Detection," *IEEE Communications Surveys & Tutorials*, 2018.

[28] A. L. Buczak and E. Guven, "A survey of data mining and machine learning methods for cyber security intrusion detection," *IEEE Communications Surveys & Tutorials*, vol. 18, no. 2, pp. 1153–1176, 2016.

[29] Y. LeCun, Y. Bengio, and G. Hinton, "Deep learning," *nature*, vol. 521, no. 7553, p. 436, 2015.

[30] D. E. Kouicem, A. Bouabdallah, and H. Lakhlef, "Internet of Things Security: a top-down survey," *Computer Networks*, 2018.

[31] F. A. Alaba, M. Othman, I. A. T. Hashem, and F. Alotaibi, "Internet of Things security: A survey," *Journal of Network and Computer Applications*, vol. 88, pp. 10–28, 2017.

[32] C. Kolias, G. Kambourakis, A. Stavrou, and J. Voas, "DDoS in the IoT: Mirai and other botnets," *Computer*, vol. 50, no. 7, pp. 80–84, 2017

[33] E. Bertino and N. Islam, "Botnets and internet of things security," *Computer*, vol. 50, no. 2, pp. 76–79, 2017.

[34] R. H. Weber, "Internet of Things–New security and privacy challenges," *Computer law & security review*, vol. 26, no. 1, pp. 23–30, 2010.

[35] M. Tavallaee, E. Bagheri, W. Lu, and A. A. Ghorbani, "A detailed analysis of the KDD CUP 99 data set," in Proc. *IEEE Int. Conf. Comput. Intell. Secur. Defense Appl.,* Jul. 2009, pp. 53–58.

10

Design a Novel IoT-Based Agriculture Automation Using Machine Learning

G. Ravi[1], Kumbala Pradeep Reddy[2], M. Mohan Rao[3], Sarangam Kodati[4], and J. Praveen Kumar[5]

[1]Department of CSE, MRCET, Hyderabad, Telangana, India
[2]Department of CSE, CMR Institute of Technology (Autonomous), Hyderabad, Telangana, India
[3]Department of CSE, Ramachandra College of Engineering, Andhra Pradesh, India
[4]Department of CSE, Teegala Krishna Reddy Engineering College, Telangana, India
[5]Department of Computer Science and Engineering, Teegala Krishna Reddy Engineering College, Telangana, India
E-mail: g.raviraja@gmail.com; pradeep529@gmail.com; mohanrao19@gmail.com; k.sarangam@gmail.com, praveen@tkrec.ac.in

Abstract

In this paper, design of a novel Internet of Things (IoT) based agriculture automation using machine learning (ML) is implemented. The major source of economy in this country is agriculture. To get better production, there should be an improvement in the technologies of IoT and ML. Because of this, there is an improvement in demand of population growth. In this, agriculture parameters are analyzed using humidity sensor, temperature sensor, and soil moisture sensor. The collected data will be sent to the web server using IoT. Machine learning algorithm will be applied for this data. Now, water motor will be on and water sprinkler will sprinkle the water and GSM message will be send to the corresponding member. At last, output is obtained. From the results, we can observe that the accuracy of the system is improved very effectively.

Keywords: Machine Learning Algorithm, IoT (Internet of Things), GSM, Soil Moisture Sensor, Temperature Sensor, Humidity Sensor.

10.1 Introduction

These days, we are encircled by a lot of "shrewd" sensors and various frameworks that provide communication associated through Internet and cloud stages; this is the Internet of Things (IoT) world view that presents trend setting innovations in all friendly and beneficial areas of the general public. This is about the overall market and organizations that contest to expand their productivity and economy by upgrading expenses, time, and assets and, simultaneously, attempting to improve the quality of administrations and the items assortment offered to clients.

The consideration toward effectiveness and beneficial enhancements is given for likewise in the rural area, where the creation elements and the sources of the board influence crop types, water systems, and water sources; keeping such creation rhythms with no programmed control is probably going to bring data squander, spoiled or deserted harvests, and clean soils and devastated soils. Creative innovations can be valuable to deal with issues like ecological maintainability, squander decrease, and soil enhancement. The social event and the examination of farming information, which incorporate various and heterogeneous factors, these are of significant premium for the chance of creating procedures which depends on the biological system and its assets (enhancement of water system and planting according to soil history and occasional cycles), the distinguishing proof of benefits and non-powerful factors, the chance of doing show case investigation comparable to the estimate of future hard-prescient data, the chance of adjusting harvests to explicit conditions, lastly the capacity to boost mechanical ventures by restricting and foreseeing equipment disappointments and substitutions. With the people blast, there is a quick expansion in the interest for food and cultivated stocks and development interaction to improve yield, cost-satisfactory, and nature of harvests/rural items being delivered with new innovation like the IoT and Artificial Intelligence. There is a need to expand yield, viability, and improved creation of land per unit territory taken under thought. It is important to accept new advancements to conquer these issues. There are different advantages related with the institution of new advances which include: expanded efficiency, legitimate harvest appropriation, crop design idea, legitimate usage of assets, for example, fertilizers and excrements utilizing the procedure of Automation and AI model.

Farming is imperative for the advancement of the world. In this, the people get benefits from farming for sure, which has made agriculture and business a critical space of study. Large farms will consistently require to show data, most particularly data for developing harvests which are not regular in their territory or culture. The normal rancher approaches unrefined wellsprings of data like TV, radio, papers, individual ranchers, government horticultural offices, ranch supply, and brokers. There is, consequently a requirement for a framework that permits ranchers process to important data. AI is among the moving advances; subsequently, there exist few innovations and frameworks that show sudden spike in demand for an AI system. As of late, a few AI frameworks in farming have been tried and made. Assessment of a couple of AI estimations' practicality in cultivating and other application regions has moreover which has been coordinated and this is because AI is a very reasonable gadget for viable use of resources, assumption, and the chiefs, which are needed in agribusiness. Artificial intelligence is the limit of an electrical getting ready structure to get data and apply that data that data.

10.2 Literature Survey

The creators Thomas Truong, Anh Dinh, and Khan Wahid, in this paper gives clarification about the gadget which gives ongoing natural information to distributed storage and an AI calculation to anticipate ecological condition for parasitic location and avoidance. In AI, calculation utilizing Support Vector Machine (SVM) was created to deal with crude information and to see the result. SVM gives result, however it is less exact than different calculations [1]. The creators Mengzhen Kang and Fei-Yue Wang, clarify in this paper that the idea of Knowledge Data Driven Model (KDDM) is utilized for new age shows agribusiness which will break the bottleneck of model application from research centre climate to genuine world [2].

The creators Yun Shi, Zhen Wang, XianfengWang, Shanwen Zhang, introduced the concept of Internet of things (IoT). Plant disease and insect pests have turned into a dilemma as it can cause significant reduction in both quality and quantity of agricultural products [1, 2]. China is one of the countries which suffer from the most serious plant disease and insect pest infection in the world. In recent decades, the loss caused by plant disease and insect pests is far more severe than that by plant fires, so plant disease and insect pests forecasting is of great significance and quite necessary.

The creators Carlos Cambra, Sandra Sendra, Jaime Lloret, Laura Garcia, clarify in this paper present the plan of a brilliant IoT correspondence framework director utilized as an ease where water system regulator. It shows how IoT, airborne pictures, and SOA can be applied to enormous and keen cultivating framework. Information is handled in shrewd cloud administration dependent on the Drools Guvnor.

[4] The creators Dr. N. Suma, Sandra Rhea Samson, S. Saranya, G. Shanmugapriya, R. Subhashri, clarify that this venture incorporates different highlights like GPS based far off controlled observing, dampness and temperature detecting, interlopers scaring, security, leaf wetness, and legitimate water system offices. It utilizes remote sensor networks for taking note of the dirt properties and natural factors constantly.

[5] The creators Xin Zhao, Haikun Wei, Chi Zhang, Kanjian Zhang, proposed a short-term wind speed for projecting a model with test determination by another dynamic learning calculation. Dynamic learning is utilized as an example choice for AI. In this investigation, dynamic learning was helpful for applications described by countless preparing tests in wind speed forecast [6]

The creators Giritharan Ravichandran, Koteeshwari RS, in this paper clarified that Artificial Neural Network is utilized which is quite possibly the best device in displaying and forecast. Feed forward Back Propagation Network is utilized together to carry out the Artificial Neural Network. The proposed framework is made as an Android Application, where the client could take care of the sources of info and get the attractive application.

[7] The creators Harshal Waghmare, Radha Kokare, in this paper support vector machine and make the choice that emotionally supportive network is utilized to recognizable proof of plant sickness through the leaf surface determination and example acknowledgment. Determination Support Systems (DSS) for horticulture depends on the innovation that can be valuable for ranchers and help to expand the farming efficiency. From this paper this comes to realize that the DSS is efficient, improve viability, and increment leader fulfillment.

[8] The creators Hemantkumar Wani, Nilima Ashtankar, clarify that AI calculation is given for the forecast of sicknesses utilizing the Bayes bit calculation. Gullible Bayes introduced the design which provide communication between ongoing information and existing informational collection. Guideless Bayes part calculation is used for the characterization of information detected from the sensors.

[9] The creators Snehal S. Dahikar, Prof. Dr. Sandeep V. Rode, Prof. Pramod Deshmukh, explain in this paper that in India cultivation is the

primary occupation. Above 70% of businesses relyon cultivation. In this paper, Artificial Neural Network innovation was utilized. The smart framework has brought fake neural network (ANN) to turn into another innovation which gives grouped answers for the pipelining issues in agribusiness investigations. This will just introduce the most regularly utilized sort of ANN, which is the feed forward back spread organization. Here, the ANN is utilized for predicting appropriate yield for specific soil and furthermore recommending legitimate compost for that crop.

10.3 Novel IoT-Based Agriculture Automation Using Machine Learning

Figure 10.1 shows the flow chart of novel IoT-based agriculture automation using ML. In this, agriculture parameters are analyzed using humidity sensor, temperature sensor, and soil moisture sensor. The collected data will be send to the web server using IoT. Machine learning algorithm will be applied for this data. Now, water motor will be on and water sprinkler will sprinkle the water and GSM message will be send to the corresponding member. At last output is obtained.

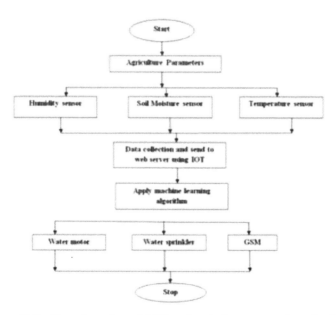

Figure 10.1 Flow chart of novel IoT-based agriculture automation using ML.

A humidity sensor is an electronic device that measures the humidity in its environment and converts its findings into a corresponding electrical signal.

Relative humidity is calculated by comparing the live humidity reading at a given temperature to the maximum amount of humidity for air at the same temperature.

The DHT22 Digital Temperature and Humidity Sensor Module AM2302 is a basic, low-cost digital temperature and humidity sensor. It uses a capacitive humidity sensor and a thermistor to measure the surrounding air and spits out a digital signal on the data pin (no analog input pins needed).

Soil moisture sensors do not measure water in the soil directly. Instead, they measure changes in some other soil property that is related to water content in a predictable way. The highest accuracy will be obtained when the soil sensor is entirely surrounded by the soil, with no gaps between the probe and the soil. A temperature sensor is an electronic gadget that actions the temperature of its current circumstance and converts the info information into electronic information to record, screen, or sign temperature changes. DS18B20 Temperature Sensor—The advanced temperature sensor DS18B20 follows a solitary wire convention and it very well may be utilized to gauge the temperature in the scope of $-67°F$ to $+257°F$ or $-55°C$ to $+125°C$ with $\pm 5\%$ accuracy. The range of obtained information from the 1-wire can go from 9-piece to 12-cycle. Since, this sensor upholds the single wire convention the managing of this should be possible through a solitary pin of microcontroller. This is a prevalent level convention, where every sensor can be set with a 64-cycle sequential code which serves to control overflowing sensors utilizing a solitary pin of the microcontroller. It is a piece of the incorporated Water System. Use a zero preceding decimal focuses: "0.25," not ".25," use "cm^3," not "cc."

Water Motor can show a wide voltage fluctuation from 180 to 240V without impacting the *engine*. This *motor* can show a wide voltage fluctuation from 180 V to 240 V without impacting the *engine*.

A water sprinkler (otherwise called a water sprinkler or just a sprinkler) is a gadget used to flood rural yields, yards, scenes, greens, and different zones. They are additionally utilized for cooling and for the control of air borne residue. GSM implies Global System for Mobile communication, which is an organization that upholds both cell and information. The equivalent goes for CDMA, which means Code Division Multiple Access. US Mobile as of now just works as a GSM organization.

10.3 Novel IoT-Based Agriculture Automation Using Machine Learning

Figure 10.2 shows the accuracy of comparison for novel IoT-based agriculture automation using ML and agriculture automation using ML. Compared to existing one, the proposed novel IoT-based agriculture using ML improves the accuracy in an effective way.

Figure 10.3 shows the performance of comparison for novel IoT-based agriculture using ML and existing agriculture automation using ML. Compared to the existing one, the proposed novel IoT-based agriculture using ML improves the performance of system in an effective way.

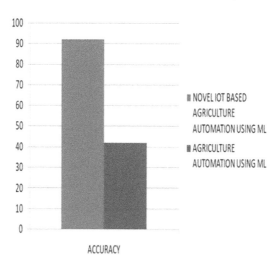

Figure 10.2 Accuracy comparison.

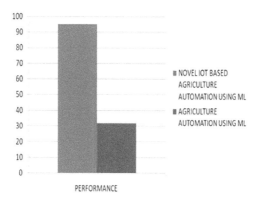

Figure 10.3 Performance comparison.

10.4 Conclusion

The proposed framework gives farming arrangement utilizing Artificial Neural Network Machine Learning calculation which is utilized for performing information forecast based on information detected by sensors. Because of utilization of IoT gadgets, framework gives computerized answers for information forecast. The delivered result will be useful for rancher to take precise choice of revenue-driven increase. The framework will give all earlier information ahead of time to the rancher for taking appropriate choice. The proposed framework will be utilized to improve the recognition of sicknesses and anticipate how the infection will spread in crop field. Same proposed framework can be stretched out to give the pesticides to anticipated sicknesses.

References

[1] Thomas Truong; Anh Dinh; Khan Wahid. An IoT environmental data collection system for fungal detection in crop fields [M]//2017 IEEE 30th Canadian Conference on Electrical and Computer Engineering (CCECE)

[2] Mengzhen Kang; Fei-Yue Wang. From Parallel Plants to Smart Plants: Intelligent Control and Management for Plant Growth [M]//2017 IEEE/CAA Journal of Automatica Sinica.

[3] Hemantkumar Wani, Nilima Ashtankar. An Appropriate Model Predicting Pest/Disease of Crops Using Machine Learning Algorithm. [C]//ICACCS 2017.

[4] Carlos Cambra, Sandra Sendra, Jaime Lloret, Laura Garcia. An IoT Service-Oriented System for Agriculture Monitoring [C]//IEEE ICC 2017 SAC Symposium Internet of Things Track.

[5] Dr. N. Suma, Sandra Rhea Samson, S. Saranya, G. Shanmugapriya, R. Subhashri. IoT Based Smart Agriculture Monitoring System. [C]//IJRITCC February 2017.

[6] Xin Zhao, Haikun Wei, Chi Zhang, Kanjian Zhang. Selective Sampling Using Active Learning for Short-term Wind Speed Prediction. [C]//IEEE 2017.

[7] Giritharan Ravichandran, Koteeshwari R S. Agricultural Crop Predictor and Advisor using ANN for Smart phones. [C]//IEEE 2016.

[8] Harshal Waghmare, Radha Kokare. Detection and Classification of Diseases of Grape Plant Using Opposite Colour Local Binary Pattern

Feature and Machine Learning for Automated Decision Support System. [C]//IC SPIN 2016.
 [9] Yun Shi, Zhen Wang, Xianfeng Wang, Shanwen Zhang. Internet of Things Application to Monitoring Plant Disease and Insect Pests [C]//International Conference on Applied Science and Engineering Innovation (ASEI 2015) 2015.
[10] Snehal S. Dahikar, Prof. Dr. Sandeep V. Rode, Prof. Pramod Deshmukh. An Artificial Neural Network Approach for Agricultural crop Yield Prediction Based on Various parameters. [C]//IJARECE 2015.

11

Building a Smart Healthcare System Using Internet of Things and Machine Learning

Shruti Kute[1], Amit Kumar Tyagi[2,3], Rohit Sahoo[4], and Shaveta Malik[5]

[1,2]School of Computing Science and Engineering, Vellore Institute of Technology, Chennai, Tamil Nadu, India
[3]Centre for Advanced Data Science, Vellore Institute of Technology, Chennai, Tamil Nadu, India
[4,5]Computer Engineering Department, Terna Engineering College, Navi-Mumbai, Maharashtra
E-mail: shrutikute99@gmail.com; amitkrtyagi025@gmail.com; rohitsahoo741@gmail.com; shavetamalik687@gmail.com

Abstract

The Internet of Things (IoT) and Machine Learning (ML) are commonly used in the field of medical diagnostics and health care to monitor a patient's condition. The IoT has been used to create systems that use the functionality of a wearable series of sensors to warn patients in the event of an abnormality. ML has assisted medical diagnosis by using models that are designed to detect anomalies within a patient's condition. Health professionals and patients benefit from the use of IoT technology because they have access to the latest up-to-date medical device settings. Furthermore, medical applications are mostly interested in IoT to minimize costs, simplify processes, and improve patient satisfaction. The increasing amount of medical information provided by the IoT is used for improving results for the patient by ML approaches in health care. These methods offer both exciting applications and significant challenges. In this paper, the developments of several applications using IoT and ML in healthcare are discussed, along with the use of 5G technology and cloud computing to manage the smart healthcare system.

Keywords: Smart healthcare Systems, Machine Learning, Internet of Things, Smart Era.

11.1 Smart Healthcare—An Introduction

The current epoch is one of digitalization. With the advancement of innovation and rational hypotheses, traditional medicine has begun to digitize, with biotechnology at its heart. Furthermore, astute patient care has ushered in a new era of data creativity. Medical care that is intelligent is not only an easy, forward-thinking move, but also a massive shift. This shift can be seen in clinical model changes, informatization implementation changes, clinical administration changes, and changes in the concept of counteraction, and therapy. These advancements are focused on addressing individual needs while increasing clinical consideration productivity, which greatly improves the clinical and wellness administration experience and points to the potential advancement direction of modern medicine. This assessment will start with the concept of savvy medical treatment, then quickly present the core developments promoting savvy medical care and explain the successes and challenges in doing so by researching the deployment status of these advancements in significant clinical areas, and finally advance the potential possibilities of smart healthcare services.

According to studies in similar areas, remote health surveillance is feasible, but the benefits it might have in different contexts are maybe more important. Remote health surveillance may be used to monitor non-critical patients at home instead of in the hospital, freeing up hospital services such as doctors and beds. It could improve access to health services for people living in rural areas or enable elderly people to stay in their homes for longer periods of time. It can essentially improve access to healthcare facilities thereby lowering the pressure on healthcare programs, and it can give individuals more control of their health at all times [1].

The IoT ensures the personalization of healthcare services by including each patient's digital identity. Many health issues have remained undetected in traditional health programs due to a shortage of ready-to-access healthcare facilities. On the other hand, IoT-based approaches that are comprehensive, noninvasive, and reliable have assisted in the rapid monitoring and review of patient data. In IoT-based healthcare, many remote machines record, store, and transmit real-time medical data to the cloud, making it much easier to collect, manage, and process vast data streams in novel ways and trigger context-dependent alerts. This innovative data collection technique allows for continuous and widespread access to medical devices via the internet from every wired IoT device.

11.2 Background Study

The concept of "Savvy Planet," suggested by IBM (Armonk, NY, USA) in 2009, resulted in shrewd medical treatment. Smart Planet is a clever system that uses sensors to gather data, transfers it over the IoT, and loops it through supercomputers and cloud computing. It has the ability to organize and integrate social structures in order to comprehend the complex and refined administration of human society. Brilliant medical services are a healthcare management system that makes use of innovation such as wearable devices, IoT, and the flexible network to efficiently access data, interface people, materials, and organizations associated with medical services, and then effectively manages and responds to clinical environment needs in a smart way [2]. Careful medical attention will promote contact between all medical meetings, ensure that the participants received the necessary administrations, assist the meetings in making informed choices and encourage a healthy portion of property. To put it plainly, keen medical care is a higher phase of data development in the clinical field.

Many participants, such as experts and patients, emergency clinics, and research institutions, make up shared care facilities. It is a natural whole that involves a variety of metrics, such as disease prediction and monitoring, analysis and treatment, clinic executives, well-being dynamics, and clinical discovery. IoT, mobile Internet, cloud computing, big data, 5G, microelectronics, and artificial intelligence, as well as modern biotechnology, form the backbone of smart medical services. WIFI, 3G/4G, Bluetooth, and NFC-RFID are only a few of the many data and exchange advancement (ICTs) that provide brilliant urban environments with more mindful, informed, and effective management dependent on human beings and city inspections [3]. These innovations are widely used in all aspects of advanced medical treatment. Patients can use wearable devices to track their health on a regular basis, search for professional assistance by menial helpers, and use far-flung homes to carry out remote administrations; experts can use a variety of intelligent clinical decision support tools to assist and enhance their decisions. Clinical records should be supervised by specialists using an integrated data stage that includes a Laboratory Information Management System, Picture Archiving and Communication Systems (PACS), Electronic Medical Record, and other systems. Careful robotics and mixed reality engineering will help doctors perform more precise surgical procedures. Radio-recurrence identifying proof (RFID) invention should be used to manage personnel materials and the gracefully chain in medical centers, with integrated administration stages gathering data and assisting dynamic. The use of portable therapeutic levels will improve patients' experiences. It is possible, from the perspective of pragmatic exploration companies, to use approaches such as AI instead of manual drug sampling and to find suitable subjects using large data sets [4]. Smart medical care can successfully reduce the cost and risk of procedures, increase the usage proficiency of clinical assets, advance trades and coordination in different districts, push the development of telemedicine and self-administration clinical consideration, and ultimately make personalized clinical administrations ubiquitous by using these advances.

11.3 Motivation of This Work

With the progress in data innovation, the concept of superb medical treatment has gradually gained traction. Smart medical care makes use of another century of data advancements, such as the IoT, big data, cloud analytics, and artificial intelligence, to transform the traditional healthcare paradigm in a big way, making medical care more efficient, beneficial, and personalized [4]. With the aim of presenting the concept of savvy medical services, we will go through the major advancements that aid savvy medical care and the existing state of savvy medical care in a few key fields in this report. We then go through the latest challenges of skilled emergency services to try to come up with solutions. Finally, we consider the potential prospects of outstanding medical treatment.

11.4 Internet of Things–Enabled Safe Smart Hospital Cabin Door Knocker

Today, every individual needs to have their own home for living with relatives. And furthermore, individuals are occupied with different exercises, for example, occupations, business, and so on. A few people select security individuals to ensure their home, a few people utilize some security components, for example, camera, lock, and so forth to ensure their homes' safety. In this paper, the creators acquainted have made sure about lock framework with secure entryways. Made sure that Smart Door framework is a mechanical or conventional entryway which is connected with electronic circuit. To get to the entryway, the clients need to utilize advanced data, for example,

11.4 Internet of Things–Enabled Safe Smart Hospital Cabin Door Knocker

a mystery code, shrewd card, or a unique mark. The term "conventional locks" is basically alluding to locks that are not mechanized and bolts that must be physically occupied with request for the locking instrument to be worked. The vast majority of these locks' components are turned on either by turning on a key, pivoting the thumb job, or squeezing a catch. In the two cases, it is physically turned on and should be physically worked to unplug the lock instrument. The most customary locks work when a key is utilized to enact the lock system, enabling it to bolt or open.

In its easiest structure, shrewd locks are programmed renditions of customary locks. IoT can be used to monitor door access remotely using a smart indicator. Much of the time, the brilliant lock will utilize the conventional lock component, yet the lock instrument can be utilized electronically or distantly. These locks vary in light of the fact that they require distinctive collaboration (between the client and the lock) than conventional locks. The name Smart Locks additionally comes from its capacity to be controlled and worked by cell phones, just as its capacity to incorporate with other brilliant gadgets. These locks permit property holders to control and secure their locks a way that conventional locks don't. On the off chance that the keen locks are working the path as it is planned to, it gives unrivaled simple entry and solace as appeared. Producers of savvy locks will in general zero in additional on the productivity and added highlights that lock brings to the table, causing them to hold back on security factors that have made locks a trademark for each home. IoT normally alludes to the association of gadgets (other than common admission, e.g., PCs, and cell phones) to the Internet. Vehicles, kitchen devices and even heart screens can be associated online things. Furthermore, as the IoT fills in the following not many years, more gadgets can be added. Secure Door Access is a straightforward advanced lock framework, that containing a 4-digit secret key put away in the program. The framework gathers the client contribution from 4 digits, and contrasts the client input and the preset secret phrase. Inside the program, if the client enters and matches put away passwords, access will be conceded (by opening the entryway). In the event that there is a crisscross between the client input and the put away secret phrase, access will be denied (by not opening the shut entryway and sending an instant message that somebody is attempting to open the entryway). Few points of interest are:

- Improve home security to maintain a strategic distance from robbery
- No requirement for key
- House proprietors will get notice in the event that anybody is getting to the entryway

- Very reasonable for the older or crippled: so, they can arrive at the entryway rapidly without requirements of other people's help.

Drawbacks

- To remember passwords
- Need electricity
- If specialized problem, can't get to the entryway.

The IoT makes use of the Internet's accessibility to have access to cloud-based resources or long-distance communication. It also relies on microcontrollers, sensors, and other circuit components to create useful devices for sending in the workplace to detect a certain signal. The information gathered by this device can be accessed by a cloud-based system, and the amount of data can also be burnt by using standard devices such as PCs, laptops, and mobile phones. G. Sowjanya and S. Nagaraju created and introduced an IoT-based door access control and safety scheme based on a biometric scanner, password, and IoT security problem [5]. M. L. R. Chandra, B. V. Kumar, and B. Suresh Babu have a basis for home security. The proposed method captures the image and transfers it to the authorized mail using the Simple Mail Transfer Protocol [6].

11.5 Smart Healthcare System Communication Protocol

Massive advancements in remote engineering and cloud computing have benefited the general population in a variety of areas. Telemedicine, or portable medical treatment, is one use that makes use of this technology, and protection is one of the key issues here. In recent years, digital media technologies have included flexible organizations, coordinated sensors, and IoT administrations. Over a long time, the topic of privacy, confidentiality, and trust has been a test in the world of IoT frameworks. Just a few projects have been suggested to support secure communication in IoT-enabled Medical Wireless Sensor Networks (MWSNs). However, existing conventions have several design flaws and are vulnerable to a variety of security threats, including sensor and client pantomime attacks. A novel engineering in MWSNs is proposed in this article, along with a suitable validated key foundation convention based on light weight Elliptic Curve Cryptography (ECC) for the engineering. The new ratification convention recognizes the security concerns that remain in current conventions. The traditional technique Burrows–Abadi–Needham (BAN) argument is used to explain the convention's precision. Further investigation also led to the conclusion that the conference is safe from actual security threats. In addition, the proposed convention is described in Verilog Hardware Description Language (HDL) and its functionalities are checked using an Altera Quartus II reproduction system for FPGA execution. Our convention's investigation and comparison to similar conventions suggest that the proposed convention is more powerful and vigorous than the existing conventions [7].

11.6 IoT-Cloud Based Smart Healthcare Data Collection System

Numerous actual articles, for example, gadgets, vehicles, buildings, and different items are utilized in customary organization framework to characterize the IoT. IoT innovation utilizes the pre-established framework of organizations to guarantee its legitimacy. Numerous mainstream brilliant gadgets are used generally, including PDAs, tablets, and sensor-prepared gadgets. The accelerometer, gyro, and proximity sensors are among the sensors used in advanced cells. Accelerometer sensors are used to measure a body's increasing speed and shift in rotational point, a closeness sensor is used to estimate the identification of nearly located objects, and GPS technology is used to estimate body positioning. RFID marks may be used to differentiate IoT papers. The number of people applying for jobs in IoT innovation is steadily growing. There is also an examination-based expectation that there will be 36.5 billion remote associations by the end of 2020, with 70 percent of remote associations containing sensor devices and 30 percent without sensors.

Various IoT projects, such as body applications, the smart house, the smart city, and smart climate ventures, all energies the use of IoT technology. Assurance and computerization ventures are used in the smart house. Contamination, temperature, tremor and tidal wave exploration, and checking projects have all been launched in order to establish the harsh atmosphere. Smart transportation, nation outskirt security, electronic administration frameworks, brilliant city gracefully chain the board, and matrix station testing are examples of keen city projects. Since IoT technology is so widely used in all aspects of life, it's no surprise that it's still widely used in the health sciences. Accordingly, unique IoT-based keen wellbeing administrations ventures are being started around the world. Different sorts of keen wellbeing administrations are being given to people in general. These may incorporate the accompanying: distant observing of patient wellbeing, understanding dealing with in a crisis, drug and routine wellbeing exam updates, far off patient medicines, and looks for the closest wellbeing assets to the patient, for example, specialists, paramedical staff, meds, rescue vehicle administrations, and numerous other wellbeing assets. Wellbeing is a significant substance for human life, and it greatly affects the economy.

Several schemes have been proposed to implement IoT technology in healthcare administrations. In the writing for IoT invention, three important ideal models have been examined: implementation, protection, and proficiency spaces. The proposed scheme focuses on the stability, proficiency, and implementation spaces of IoT advancement in health care. With IoT creativity, this scheme provides a general strategy and implementation approach. It addresses all of the innovation's commonsense application problems (i.e., correspondence substances, correspondence advances, equipment structure, information stockpiling, information stream, and access instruments). Zhao et al. [8] suggested a structure that observed elderly people from afar. They focused on the IoT innovation application space. Their model catered to the needs of senior citizens. In their proposed model, they used AI procedures. Basanta et al. suggested another structure, Help to You (H2U), to screen elderly people and their health status [9]. Swiatek and Rucinski [10] proposed the importance of distant checking for patients in the application worldview. Their proposed model emphasized an appropriated structure for delivering smart health administrations. They have looked after the creative and commercial aspects of e-healthcare services. Yang et al [11] combined the traditional concept of a hospital package with excellent health administrations. Medical forms of assistance are provided using iMedBox, iMedPack, and Bio-Patch. Fan et al. [12] suggested a productive asset improved reconstruction framework in IoT conditions. In their proposed model, semantic evidence is used to proficiently understand the therapeutic assets of a keen health system. Ding et al. [13] suggested a scheme to keep the electronic Healthcare record and area data confidential, which took aim at the security worldview of IoT invention. Regardless, their main focus was on the protection of individual areas, individual IDs, identifiable evidence of queries, and individual electronic health records. Guigang Zhang, Chao Li, Yong Zhang, and colleagues [14] suggested that clients/patients' health data be verified by using explicit principles to calculate the massive IoT clinical sensor data.

11.7 Use of Machine Learning in Different Fields of Medical Science

With regards to viability of AI, more information quite often yields better outcomes and the medical services area is perched on an information goldmine.

McKinsey gauges that large information and AI in pharma and medication could produce an estimation of up to $100B yearly, in light of better dynamic, advanced development, improved productivity of examination/clinical preliminaries, and new apparatus creation for doctors, purchasers, safety net providers, and controllers. Where does this information come from? In the event that we could take a gander at named information streams, we may see innovative work (R&D); doctors and centers; patients; parental figures; and so forth the variety of (as of now) divergent beginnings is important for the issue in synchronizing this data and utilizing it to improve medical care foundation and therapies. Consequently, the present-day center issue at the convergence of AI and medical care: discovering approaches to viably gather and use bunches of various sorts of information for better investigation, avoidance, and therapy of people. Thriving uses of ML in pharma and medication are flickers of an expected future in which synchronicity of information, examination, and advancement are a regular reality. In this segment, we give breakdown of a few of the spearheading utilizations of AI in pharma and territories for proceeding with development. ML holds much promise for accurately identifying a variety of findings in pictures. While the mainstream press has used computers to replace radiologists, the most exciting thing maybe that ML appears to be able to find features in images not visible to people [15].

11.8 Illness Identification/Diagnosis

ML used to differentiate between illnesses and diagnose diseases is at the cutting edge of medical science. More than 800 drugs and antibodies to treat malignant growth are provisional, according to a 2015 study by Manufacturers of America and Pharmaceutical Research. Although this is energizing, Knight Institute Researcher, Jeff Tyner said in a meeting with Bloomberg Technology that while it is energizing, it also poses the challenge of figuring out how to deal with all subsequent data. "That is the place where the possibility of a scholar working with data researchers and computation lists is so significant," said Tyner. Surprisingly, significant players were among the first to jump on board with the innovation [17]. Ebb and Flow Research ventures in progress incorporate measurement preliminaries for intravenous tumor therapy and location and the executives of prostate malignant growth.

11.8.1 Discovery of Drug & Manufacturing

AI is helpful in discovery of drug. ML applies in early stage of the discovery of the drugs. As next-generation sequencing and precision medicine which can support in finding alternative paths for treatment of multifactorial diseases in the R & D technologies [18]. Now, unsupervised learning algorithms are being used which is able to identify the pattern without using the prediction. AI technologies are used in treatment of lot of diseases such as cancer and moreover in personalizing of drugs.

11.8.2 Diagnosis of Medical Imaging

Through the AI and ML it is easy to diagnose the disease in a short period of time. Automatic detection of the abnormalities is helpful in taking quick decisions related to any diseases. It also controlled to lessen diagnostic errors. For example, sometimes type of fracture is difficult to detect through standard images but AI tools help in detecting minor variations also in image that is useful in the surgery.

11.8.3 Clinical Trial

Machine Learning is helpful in clinical trials. Through the ML prediction analysis we can identify the candidate for the clinical trial. Also, Machine Learning is more helpful in tele-medicine or healthcare sector to identify reasons behind the causes/effects of diseases upon humans. It is also helpful in diagnosing the disease of the patient through previous history by prediction analysis. ML also helps in remotely monitoring the patients. There are a number of applications of ML that are useful in increasing the efficiency of the clinical trial, for example,, evaluating the best size of a sample, helpful in reducing the error, duplicate data in EMR (Electronic Medical Record).

11.8.4 Epidemic Outbreak Prediction

AI and ML are helpful in reduction of the outbreak disease. SVM (Support vector machine) and Neural Networks are helpful in predicting the factors related to malaria such as temperature and total number of positive cases. ML algorithms are helpful in predicting the disease alert through the previous data prediction analysis.

11.8.5 Robotic Surgery

It is the latest field in the healthcare industry. ML applications are used in supporting robotic surgery. It reduces the procedure time and also minimizes the fatigue of the surgeon. ML techniques are also very helpful in decreasing the utilization of the resources especially at the time of pandemic even when the number of patients is higher than usual.

11.8.6 Smart Health Record

Most of the unstructured data stored in Electronic Health record system and that is unreadable or lot of the data stored in the text files and past data are not analyzed by the humans. Sometimes, the data is unreadable due to ambiguity, jargons, and lack of uniformity. So, ML is accommodating in covert that data into the useful data through Natural Language processing (NLP). NLP is also used in social media through sentimental analysis to read the important charts of patients, it helps to extract the important data.

11.9 Challenge'S Faced Towards 5G With IOT And Machine Learning Technique

Creating advances, for example, 5G organizing, IoT, and AI are empowering everything from self-driving vehicles to common language acknowledgment. In any case, these promising innovations are additionally making new network protection challenges for IT security experts.

"There are a few advancing assault vectors, however the three that truly keep me up around evening time are 5G networks, the internet of things, and AI or AI," said Paul Mazzucco, Chief Security Office for TierPoint. Mazzucco clarified how these arising three assault vectors will affect network safety in an ongoing webcast, Bots evolve to challenge security in a 5G World.

11.9.1 5G and IoT Empower More Assault Vectors

5G networks are turning out in most significant urban communities, and Verizon predicts that a big part of the United States will before long be utilizing 5G. Other than quicker speeds, 5G offers the capacity to characterize

Technical Scenarios	Key Issues
High-capacity hot-spot	Traffic Volume density: Tens of Tbps/Km^2
	User experienced data rate: 1 Gbps
	Peak data rate: Tens of Gbps
Low-power massive	Connection density: 106/km^2
	Low power consumption and low cost
Seamless wide-area coverage	100 Mbps user experienced data rate
Low latency high-reliability	Air interface latency: 1 ms
	End-to-end latency: ms level
	Reliability: nearly 100%

Figure 11.1 Main problems in 5G.

use-case based virtual organizations made to meet the scope of necessities of IoT gadgets. For instance, a self-driving vehicle may send and get volumes of information for examination, though an ecological sensor may communicate just temperature and moistness changes. For constant surveillance of chronic patients, a smartphone utilizing the next-generation cellular data network, namely 5G, and an IoT-based solution has been suggested in [20]. Also, few problems with respect to 5G has been listed in Figure 11.1. 5G empowers many tweaked networks over existing framework. That ability is extending the potential use cases for IoT. There are now an expected 27 billion IoT gadgets, in businesses going from assembling, transportation, and broadcast communications to medical services, places of business, and purchaser homes. Estimates recommend that the quantity of IoT gadgets will arrive at 50 billion by 2030. Tragically, numerous IoT gadgets have frail security, made more fragile via indiscreet clients who regularly disregard to change the manufacturing plant default passwords on their gadgets. The developing number of unstable IoT gadgets implies more open doors for cyber attackers who can utilize them to invade corporate organizations and make botnets to dispatch Denial of Service (DoS) assaults.

Nor is 5G without its own security weaknesses. To make separate virtual organizations, 5G utilizes Software-Defined Networking (SDN). These numerous product-based organization "cuts" offer cyberattacks more focuses to hack and expands the chances they can recognize singular sorts of traffic. Moreover, 5G utilizes short-separation transmissions, making it subject to various, little cell towers situated close to sending associations. That, once more, grows the quantity of expected targets and assists programmers with pinpointing which pinnacles are destined to communicate a specific association's traffic. "A programmer can plunk down close to the framework, track down those parcels, and assault specific ones," said Mazzucco, referring to medical clinics with their volumes of by and by recognizable data (PII) as high-esteem targets.

11.9.2 Smarter Bots Can Likewise Misuse These Assault Vectors

AI is making everything more astute, so it's nothing unexpected that AI is additionally helping programmers make more brilliant wrongdoing bots. Bots are pieces of code that do fundamental undertakings. Dissimilar to botnets which are organizations of IoT gadgets contaminated with bot malware, singular bots are programs that do things like, search the web, scratch data off sites, and retweet news things via online media. An arising fourth era of bots is utilizing AI to impersonate human conduct—and doing it all around ok that numerous security arrangements neglect to identify them. "Programmers are utilizing artificial intelligence and AI to be more viable," noted Mazzucco, noticing that assessed 20% of Internet traffic is brought about by noxious bots. For example, 4G bots can be daisy-binded to perform complex, robotized assaults which are difficult to recognize in light of the fact that they so consummately execute human conduct—directly down to moving the mouse pointer in an arbitrary example much like a human examining a page. In "fourth era bots realize how long to press on symbols to emulate people and will even incorrectly spell words deliberately to look more legitimate," notes Mazzucco. "We call it conduct capturing utilizing progressed AI."

AI-controlled bots can likewise insect conditions to discover what security conventions are set up, what the best assault vector is, and by what means should the assault be organized for ideal achievement. There is a positive side to this. The product-based nature of 5G should make it simpler to screen traffic for explicit dangers and adjust security arrangements for various sorts of traffic. Moreover, AI and mechanization will likewise assist with supporting 5G security. For instance, robotized examination of traffic designs utilizing AI can realize what is ordinary versus strange conduct for explicit gadgets, applications, and end-clients, and spot potential assault designs. Relapse testing can take a gander at designs throughout extensive stretches of time, for example, 90 days past, and distinguish related abnormalities separated days or weeks separated.

AI is additionally helping network protection designers make fourth era "white cap" bots to search out wrongdoing bots and stop them before they can do harm. For instance, AI-based innovations, for example, goal-based profound conduct examination, are being created to all the more precisely and rapidly distinguish terrible bot practices. In last, all readers are requested to read our papers [21–29] to know more about IoT, IoT-based Healthcare, Machine Learning role in Healthcare, Deep Learning, etc.

11.10 Future Possibility of Smart Healthcare With Internet of Things

IoT and AI together are to play key role in future smart hospitals. The blend of AI IoT will uphold the advancement of shrewd medical clinics and fuel the progressing development of large information investigation. IoT is required to consolidate with the intensity of AI, blockchain, and other arising advancements to make the "brilliant emergency clinics" of things to come, as per another report by Frost and Sullivan. The IoT—additionally usually referred to in the medical care industry as the Internet of Medical Things (IoMT)—comprises of all clinical gadgets, quiet checking apparatuses, wearables, and different sensors that can impart signs to different gadgets by means of the internet.

- These apparatuses create enormous measures of information that must be put away, incorporated, and dissected to produce noteworthy experiences for ongoing infection the executives and intense patient consideration needs.
- IoT information is a significant expansion to other clinical information sources, for example, the Electronic Health Record (EHR), that permits suppliers to screen patients on a continuous premise or foresee changes in a person's wellbeing status.
- "Raising interest for distant patient observing, alongside the presentation of cutting-edge cell phones, versatile applications, wellness gadgets, and progressed medical clinic framework, are making way for setting up keen emergency clinics everywhere on the world," says the report.
- Prescient examination techniques are starting to depend on the accessibility of information from wearables and IoT gadgets both inside and outside of the medical clinic.
- Anticipating understanding decay or contamination in the inpatient setting requires nonstop input from bedside gadgets, while home checking apparatuses, for example, Bluetooth-empowered circulatory strain sleeve, scales, and pill containers can keep patients' disciple to persistent sickness the executives' conventions outside of the facility.

As indicated by an ongoing investigation by Deloitte, more than 66% of clinical gadgets will be associated with the internet by 2023, contrasted with only 48% of gadgets in 2018. The uptick in associated gadgets will prompt the accessibility of more information for investigation, which will thusly require novel strategies for extricating significance from crude datasets. An artificial intelligence and AI procedure are obviously adjusted to overseeing and examining nonstop information streams in enormous sums, says Frost and Sullivan, and will be basic for guaranteeing that significant bits of knowledge are introduced to suppliers without over-burdening their work processes. "Sensors, artificial intelligence, big data analytics, and blockchain are vital technologies for IoMT as they provide multiple benefits to patients and facilities alike," said Varun Babu, Senior Research Analyst, TechVision. "For example, they help with the conveyance of focused and customized medication while at the same time guaranteeing consistent correspondence and high efficiency inside savvy clinics." The possibility to improve productivity, draw in patients constantly, and stretch out beyond unfriendly occasions has made a huge business open door for gadget makers, programming sellers, and examination designers, adds a different report by MarketersMedia.

Presently, the worldwide IoT market is esteemed at $20.59 billion, and is foreseen to develop at a 25.2% Compound Annual Growth Rate (CAGR) until 2023 to reach $63.43 billion. The market incorporates implantable apparatuses, for example, heart gadgets, just as internet-associated ventilators, imaging frameworks, indispensable signs screens, respiratory gadgets, implantation siphons, and sedation machines, MarketersMedia says. Ice and Sullivan likewise foresees that arising classifications of IoT gadgets, including cement skin sensors, will add to the monetary chance, while creating innovations, for example, blockchain, will improve the security, interoperability, and investigation capability of these apparatuses. To succeed, suppliers and designers should work together on making and conveying information guidelines and shared conventions to guarantee the consistent trade of information across dissimilar frameworks. "The primary goal of IoMT is to kill pointless data inside the clinical framework so that specialists can zero in on conclusions and treatment," said Babu. "Since it is an arising innovation, innovation engineers need to offer state sanctioned testing conventions with the goal that they can persuade emergency clinics regarding their wellbeing and viability and capitalize on their gigantic potential."

11.11 Conclusion and Future Scope

Throughout the long term, mechanical advancements in the medical services industry have led to improved diagnostics and expanded the nature of patient consideration. From headways in exploratory prosthetic innovation to the utilization of artificial intelligence in retinopathy conclusion, medical care is creating at a fast movement. While the Internet of Things (IoT) has just been a vital part of the medical care development, late occasions have featured its current and future potential in medical care innovation. Note that the presentation of smart pills can improve treatment and care in near future. One key progression in clinical treatment is the presentation of "savvy pills" that contain minute sensors to communicate persistent information once gulped, inserting itself in the stomach lining. In the territory of examination improvement, Proteus Discover's brilliant pill can gauge how viable medicines are all in all. They help tell specialists when patients have taken their medication, in light of the fact that for specific conditions like malignancy, sadness, and schizophrenia, patients can battle to adhere to a timetable. It likewise monitors their action levels, with the goal that doctors can improve proposals for their wellbeing. With the developing number of cell phone clients worldwide and the presentation of 5G, the general population is getting on to how simple online channels are for clinical purposes, just as retail use. Moreover, measures to ensure medical care experts, for example, distantly worked disinfecting instruments are bound to be sent later on. Finally, with the inescapable effect of respiratory sicknesses, for example, COVID-19, the contribution of clinical trackers and wellness gadgets are significant in giving early recognition and therapy.

References

[1] S.B. Baker, W. Xiang and I. Atkinson, "Internet of Things for Smart Healthcare: Technologies, Challenges, and Opportunities," in IEEE Access, vol. 5, pp. 26521–26544, 2017, doi: 10.1109/ACCESS.2017.2775180.

[2] M.N. Mohammed, S.F. Desyansah, S. Al-Zubaidi, E. Yusuf. "An internet of things-based smart homes and healthcare monitoring and management system: Review", Journal of Physics: Conference Series, 2020.

[3] Jia-Li Hou, Kuo-Hui Yeh. "Novel Authentication Schemes for IoT Based Healthcare Systems", International Journal of Distributed Sensor Networks, 2015.

[4] Priya Dalal, Gaurav Aggarwal, Sanjay Tejasvee. "Preeminent Development Boards to Design Sustainable Integrated Model of a Smart Healthcare System under IoT", IOP Conference Series: Materials Science and Engineering, 2021.

[5] Sowjanya, G., Nagaraju, S.: Design and implementation of door access control and security system based on IOT. In: 2016 International Conference on Inventive Computation Technologies (ICICT), Coimbatore, 1–4 (2016).

[6] Chandra, M.R., Kumar and B.V., SureshBabu, B.: IoT enabled home with smart security. In: 2017 International Conference on Energy, Communication, Data Analytics and Soft Computing (ICECDS), Chennai, 1193–1197 (2017).

[7] Venkatasamy Sureshkumar, Ruhul Amin, V.R. Vijaykumar, S. Raja Sekar, "Robust secure communication protocol for smart healthcare

system with FPGA implementation, Future Generation Computer Systems", Volume 100, 2019, Pages 938–951, ISSN 0167-739X.
 [8] W. Zhao, C. Wang, Y. Nakahira, Medical Application on Internet of Things, 2011.
 [9] H. Basanta, Y.P. Huang, T.T. Lee, Intuitive IoT-based H2U healthcare system for elderly people, in: 2016 IEEE 13th International Conference on Networking, Sensing, and Control (ICNSC), IEEE, 2016, April, pp. 1–6.
[10] P. Swiatek, A. Rucinski, IoT as a service system for eHealth, in: 2013 IEEE 15th International Conference on e-Health Networking, Applications & Services (Healthcom), IEEE, 2013, October, pp. 81–84.
[11] G. Yang, L. Xie, M. M`antysalo, X. Zhou, Z. Pang, L. Da Xu, S. Kao-Walter, Q. Chen, L.R. Zheng, A health-IoT platform based on the integration of intelligent packaging, unobtrusive bio-sensor, and intelligent medicine box, IEEE Trans. Ind. Inform. 10 (4) (2014) 2180–2191.
[12] Y.J. Fan, Y.H. Yin, L. Da Xu, Y. Zeng, F. Wu, IoT-based smart rehabilitation system, IEEE Trans. Ind. Inform. 10 (2) (2014) 1568–1577.
[13] D. Ding, M. Conti, A. Solanas, A smart health application and its related privacy issues, in: 2016 Smart City Security and Privacy Workshop (SCSP-W), IEEE, 2016, April, pp. 1–5.
[14] Guigang Zhang, Chao Li, Yong Zhang, Chunxiao Xing and Jijiang Yang, "SemanMedical: A Kind of Semantic Medical Monitoring System Model Based on the IoT Sensors" , IEEE 14th International conference e-Health Networking, Applications and Services Healthcom), 2012,pp 238-243.
[15] Erickson, B.J. Machine Learning: Discovering the Future of Medical Imaging. *J Digit Imaging* **30,** 391 (2017). https://doi.org/10.1007/s10278-017-9994-1.
[16] Jabbar, M.A., Shirina Samreen, and Rajanikanth Aluvalu. "The future of health care: Machine learning." *Int J Eng Technol.* 7, no. 4 (2018): 23-5.
[17] Pillai R., Oza P., Sharma P. (2020) Review of Machine Learning Techniques in Health Care. In: Singh P., Kar A., Singh Y., Kolekar M., Tanwar S. (eds) Proceedings of ICRIC 2019. Lecture Notes in Electrical Engineering, vol 597. Springer, Cham. https://doi.org/10.1007/978-3-030-29407-6_9
[18] A. Linn, How Microsoft computer scientists and researchers are working to 'solve' cancer, 2018, https://news.microsoft.com/stories/computingcancer/,accessed:2018-04-10.

[19] Q. Feng, D. He, H. Wang, L. Zhou and K. R. Choo, "Lightweight Collaborative Authentication with Key Protection for Smart Electronic Health Record System," in *IEEE Sensors Journal*, vol. 20, no. 4, pp. 2181–2196, 15 Feb. 15, 2020, doi: 10.1109/JSEN.2019.2949717.
[20] J. Lloret, L. Parra, M. Taha and J. Tomás, "An architecture and protocol for smart continuous eHealth monitoring using 5G", *Comput. Netw.*, vol. 129, pp. 340–351, Dec. 2017.
[21] B. Gudeti, S. Mishra, S. Malik, T. F. Fernandez, A. K. Tyagi and S. Kumari, "A Novel Approach to Predict Chronic Kidney Disease using Machine Learning Algorithms," 2020 4th International Conference on Electronics, Communication and Aerospace Technology (ICECA), Coimbatore, 2020, pp. 1630–1635, doi: 10.1109/ICECA49313.2020.9297392.
[22] Nair M.M., Kumari S., Tyagi A.K., Sravanthi K. (2021) Deep Learning for Medical Image Recognition: Open Issues and a Way to Forward. In: Goyal D., Gupta A.K., Piuri V., Ganzha M., Paprzycki M. (eds) Proceedings of the Second International Conference on Information Management and Machine Intelligence. Lecture Notes in Networks and Systems, vol. 166. Springer, Singapore. https://doi.org/10.1007/978-981-15-9689-6_38
[23] Akshara Pramod, Harsh Sankar Naicker, Amit Kumar Tyagi, "Machine Learning and Deep Learning: Open Issues and Future Research Directions for Next Ten Years", Book: Computational Analysis and Understanding of Deep Learning for Medical Care: Principles, Methods, and Applications, 2020, Wiley Scrivener, 2020.
[24] Tyagi, Amit Kumar and G, Rekha, Machine Learning with Big Data (March 20, 2019). Proceedings of International Conference on Sustainable Computing in Science, Technology and Management (SUSCOM), Amity University Rajasthan, Jaipur - India, February 26–28, 2019.
[25] Amit Kumar Tyagi, Poonam Chahal, "Artificial Intelligence and Machine Learning Algorithms", Book: Challenges and Applications for Implementing Machine Learning in Computer Vision, IGI Global, 2020.DOI: 10.4018/978-1-7998-0182-5.ch008
[26] Kumari S., Vani V., Malik S., Tyagi A.K., Reddy S. (2021) Analysis of Text Mining Tools in Disease Prediction. In: Abraham A., Hanne T., Castillo O., Gandhi N., Nogueira Rios T., Hong TP. (eds) Hybrid Intelligent Systems. HIS 2020. Advances in Intelligent Systems and Computing, vol 1375. Springer, Cham. https://doi.org/10.1007/978-3-030-73050-5_55

[27] Amit Kumar Tyagi, G. Rekha , "Challenges of Applying Deep Learning in Real-World Applications", Book: Challenges and Applications for Implementing Machine Learning in Computer Vision, IGI Global 2020, p. 92–118. DOI: 10.4018/978-1-7998-0182-5.ch004

[28] Varsha R., Nair S.M., Tyagi A.K., Aswathy S.U., RadhaKrishnan R. (2021) The Future with Advanced Analytics: A Sequential Analysis of the Disruptive Technology's Scope. In: Abraham A., Hanne T., Castillo O., Gandhi N., Nogueira Rios T., Hong TP. (eds) Hybrid Intelligent Systems. HIS 2020. Advances in Intelligent Systems and Computing, vol. 1375. Springer, Cham. https://doi.org/10.1007/978-3-030-73050-5_56

[29] Tyagi, Amit Kumar; Nair, Meghna Manoj; Niladhuri, Sreenath; Abraham, Ajith, "Security, Privacy Research issues in Various Computing Platforms: A Survey and the Road Ahead", Journal of Information Assurance & Security. 2020, Vol. 15 Issue 1, p. 1–16. 16 p.

12

Research Issues and Future Research Directions Toward Smart Healthcare Using Internet of Things and Machine Learning

Shruti Kute[1], Amit Kumar Tyagi[2,3], and Meghna Manoj Nair[4]

[1,2,4]School of Computing Science and Engineering, Vellore Institute of Technology, Chennai, Tamil Nadu, India
[3]Centre for Advanced Data Science, Vellore Institute of Technology, Chennai, Tamil Nadu, India
E-mail: shrutikute99@gmail.com; amitkrtyagi025@gmail.com; mnairmeghna@gmail.com

Abstract

Machine Learning (ML) and Internet of Things (IoT) are two of the highly discussed and extensively used research areas in the current world. In highly essential and sensitive arenas of healthcare, a perfect blend of these two components can pave way to striking innovations which would have the potential to cure numerous diseases and health-related issues. One of the main reasons for combining IoT and subsets of AI is because they are complementary to each other from a plethora of perspectives. IoT can be easily used to handle and deal with voluminous amounts of data flow and processing while the ML algorithms can be used to extract useful and necessary information, for training data models, and so on. Furthermore, this unique combination of two major technical concepts would improve and optimize the efficiency of operations in the medical field. It would significantly reduce the load and the efforts to be put in by the medical fraternity while they can indulge in exploring deeper into the dwellings of medical sciences in curing diseases. Moreover, IoT connects the different devices as per these needs and instantiates a smooth flow of interaction between the interconnected gadgets and devices. This paper mainly talks about the role of IoT and ML in healthcare and elucidates the necessities of wearable systems and communications standards.

Keywords: Smart Devices, Internet Connected Things, Machine Learning, AI–IoT-based Future Healthcare.

12.1 Introduction

The current scenario is extremely different when compared to that of the past years. Studies by World Wellbeing Association have shown a plethora of cases to support the same. It also predicts that at this rate, 2030 is likely to be extended by around three more years. However, the possibility of future varies completely for females in the society, for instance by around 8 years for ladies in Poland. This often leads to adaptability changes and issues encompassed by other medical concerns [1]. This calls for frameworks and systems that would motivate the existence of such techniques without paying for the cost of security and privacy. This can be made possible by evolving methodologies which possess a self-ruling framework, capable of monitoring a person's position and actions and responses.

Exploration in related fields has demonstrated that far off health monitoring is conceivable, yet maybe more significant are the benefits it could give in various settings. Distant health monitoring could be utilized to screen non-basic patients at home as opposed to in medical clinic, lessening strain on hospital resources, for example, specialists and beds. It very well may be utilized to provide better admittance to medical services for those living in rustic areas, or to empower old individuals to live freely at home for longer. Basically, it can improve admittance to healthcare resources while diminishing strain on medical services frameworks, and can give individuals better authority over their own wellbeing at all times [2].

12.2 Background Work

A plethora of surveys, researches, and contributions have been made in the field catering to Internet of Things (IoT) and Machine Learning (ML) especially with regards to the medical arena. Authors R. Prasad and H. Kobayashi have been trying to bring about an enhancement in equipment depiction language framework through a profitable means. They also plan on highlighting the nine-venture multi-technique configuration measure model which consists of architectural intricacies, detailing, segment recreation, etc. [3]. Their answer empowers the reduction of the time needed for displaying and reproduction-related exercises by 31% and 16%, separately, contrasted with the traditional equipment depiction language-based plan [4].

The proposal put forth by S. A. Megel et al. includes three major phases: prerequisites, determination, and utilization. In the first stage, the architects must narrow down the critical problem/concern and highlight a chart with the design of the framework. In the second stage, the proposal chart is polished and converted into the substance flowchart, which would be efficiently implementable into the constructed framework in the final stage. Also, after every DM stage, the approval and confirmation should be done to guarantee that the key ideas would have been met [3].

To improve the efficiency of the perplexing gadgets framework plan, H. Eskelinen proposes to apply two surveys to the customary four-stage hardware framework plan, which are: framework plan, gadgets plan, mechanical plan, and plan for assembling. Those surveys are utilized to shape necessities arrangements of electronic framework segments.

F. Wang and M. J. Hannafin express that the plan-based exploration should be "even minded, grounded, intelligent, iterative, adaptable, integrative, and relevant." In light of this presumption they structured nine standards of the plan-based exploration: uphold plan with research from the beginning; set reasonable objectives for hypothesis advancement and build up an underlying arrangement; direct the examination in delegate genuine settings; work together intimately with members; execute research techniques deliberately and intentionally; dissect information quickly, persistently, and reflectively; refine plans consistently; report logical impacts with plan standards; and approve the generalizability of the plan.

A. Saini and P. Yammiyavar picked the client as the point of convergence of the plan of m-wellbeing framework. They utilize the item situated framework plan approach, common for programming advancement, and afterward study associations and connections between the framework prerequisites and the segments of the client's necessities and objectives. Client driven plan turns out to be particularly helpful in wellbeing applications, where the partners and various types of clients may communicate various prerequisites and requirements.

The recommended DM approach of M. Ahmad thinks about five plan angles: the objective field disappointment rate, expected use climate, expected climate use conditions, expected walled in area use conditions, and expected item inner conditions. The strategy is pertinent to appraise the objective's lifetime in the IoT. It utilizes the probabilistic methodology for assessing equipment dependability with given unsure use conditions while thinking about generally speaking framework unwavering quality.

Arising advancements make new chances, and the hearty observing of people or things, the same, in indoor and open air conditions, is the fate important to numerous logical and modern applications, where one of the most significant is the medical care area. Notwithstanding, the directed review uncovers that plan philosophies, regardless of their viability, have not yet been of extraordinary interest to creators in the field of medical care data frameworks in IoT. The arising medical care applications are conceivable because of the advancement in Miniature Electro-Mechanical Systems (MEMS), which empower the mix of different gadgets like actuators, sensor hubs, or mobiles.

WSNs are widely being used in medical technologies mainly because of their applications and variation. C. Rotariu and V. Manta have put forward a WSN in order to monitor the pulse and oxygen levels of a patient. W. Y. Chung, S. C. Lee and S. H. Toh implant Electrocardiography (ECG) and circulatory strain sensors into a mobile phone [3]. The remote body territory network is an illustration of an appropriate way to deal with the IoT medical care worldview. S. - L. Tan, J. Garcia-Guzman, and F. Manor Lopez use Wi-Fi innovation to send information about the pulse, internal heat level, and oxygen immersion to the base station. J. Wannenburg and R. Malekianc apply Bluetooth innovation and a cell phone for checking the patient's wellbeing boundaries [3].

In IoT medical services applications, one of the frequently observed issues is the restriction of patient or gear. For this reason, contingent upon the application, different techniques and innovations are utilized. Various methodologies depend on Received Signal Strength (RSS). M. Shchekotov utilizes RSS estimations from a few realized Wi-Fi passages guaranteeing the confinement precision at a four room-zone level on a solitary floor of a structure. To confine a resource in the medical services climate, the creators utilize the current framework of the Wireless Local Area Network (WLAN), broadened just with six passageway reference points. In view of Wi-Fi RSS estimations and little Wi-Fi labels, they can limit the resources like wheelchairs, beds, and so forth, with a precision of around 2 m in the clinic center climate of 63 m × 46 m size. W. H. Chen et al. use RFID RSS estimations of the reference and observed labels to assess the cost work comprising of the dissimilarity and likeness of RSS among checked and reference labels. Thusly, the three ideal reference labels are found, and the situation of the observed tag is resolved as the focal point of mass of the triangle, which they structure. The normal confinement mistake of a patient or resource in a 5 m × 10 m medical care climate is about 0.74 m. F. Palumbo et al. propose the stigmergy approach joined with RSS estimations of Bluetooth Low Energy (BLE). Their methodology brings about a limitation blunder of less than 1.8 m in 75% cases in a 6 m × 6 m outfitted office. J. Wyffels et al. propose a medical care devoted indoor confinement calculation dependent on BLE RSS estimations and least squares-uphold vector machine, coming about at the four room-zone level limitation precision. The creators of this proposal would focus on the patients' confinement and follow up on the nursing foundation climate. They use RSS estimations of the ZigBee standard and a molecule channel. Therefore, they accomplished a normal confinement blunder of under 2 m in 80% of cases [3].

Various calculations and strategies can be utilized to improve the restriction precision. The creators utilize the Radio Frequency IDentification (RFID) fingerprints technique and the counterfeit neural organization, which empowers a 3D confinement precision of around 70 cm inside a room-sized climate. An alternate way to deal with the indoor restriction issue is appeared in where the creators utilized fingerprints of Wi-Fi and barometric strain to limit an objective with the floor exactness of a six-story building.

An intriguing arrangement of the Person on foot Dead Retribution (PDR) calculation is introduced by Kang and Han. They use information from off-the-rack three-hub gyrator, magnetometer, and accelerometer cell phone sensors in an in-building climate. The proposed technique guarantees the mean limitation precision of 1.35 m with the greatest confinement mistake of 1.62 m. The creators of utilization information from the accelerometer, magnetometer, and whirligig to perceive an individual's stance and to distinguish the tumbling of the individual [3].

Data about the situation of an observed individual or hardware is significant for limitation, yet additionally it very well may be utilized for patient's conduct acknowledgment. This is particularly valuable while observing the old living alone or an individual at the primary phases of dementia. For this reason, L. Wang et al. apply coin-sized RFID per users on two hands of a patient and one accelerometer on the patient's abdomen. Utilizing this set, alongside a detached RFID tag, they can perceive 25 distinct exercises of the directed individual. H. Martin et al. can perceive an individual's exercises and body position by methods of Google Nexus S inserted sensors like the magnetometer, whirligig, accelerometer, light and closeness sensors, and a fluffy classifier.

A large portion of the referenced checking arrangements have the basic downside of being committed just to indoor climate applications. On account of an open air medical services observing reason, a large portion of the empower arrangements apply the Global Positioning System (GPS), which in the in-city climate gives restriction precision of around 6 m. Ch. Wu et al. join GPS information with whirligig and accelerometer information utilizing the dead retribution calculation, which brings about an improvement of the in-city restriction exactness up to 4 m. For open air conduct acknowledgment, L. Sun et al. apply the versatile implanted accelerometer and Support Vector Machine (SVM)- based classifier, to perceive exercises like bicycling, running, and strolling [3].

The referenced checking arrangements are committed only to only one, an indoor or open air, climate. A multi-climate restriction arrangement was proposed by Millner et al, wherein the creators, utilizing the Symeo neighborhood situating RADAR, can restrict creatures with an exactness of 0.5 m in 75% of cases for both indoor and outside conditions; nonetheless, the significant limitation of the framework is its pertinence in a climate with low multi-path twists. J. Gonzalez et al. consolidate Super Wide Band (UWB) and GPS innovations and a molecule channel to confine a robot in the indoor and open air conditions with a restriction exactness of around 2 m.

These multinatural arrangements, hence, are hard to execute in medical services applications because of the size of the gadgets utilized. A restriction framework moderately handily executed in medical services, both indoor and open-air conditions, is introduced in. It depends on RSS estimations in a ZigBee network. The significant disadvantage of this arrangement is countless required reference hubs with known positions and the most extreme good ways from the reference hub of 15 m, which impressively decreases the materialness of the framework from the enormous open-air climate.

One of the best ways to handle multi-ecological patient monitoring architecture as proposed by R Tabish et al. is to make use of a technique that highlights an observing arrangement of the temperature of the patient along with the ECG which is dependent on 3G/Wi-Fi for gathering the data and information from the sensors. While the checked individual happens in an indoor climate, the framework utilizes nearby Wi-Fi for sensors' information move, and on account of the outside climate, the 3G/4G innovation is applied. The disadvantage of this arrangement is a predetermined number of observed imperative boundaries [3].

12.3 Healthcare and Internet of Things

As per the previous section, it can be analyzed that IoT is indeed an integral component in the field of medicine. The use of IoT in healthcare has proliferated over the years to a great extent. Concurrently, we can observe how the different medical care IoT use cases are getting a push and move one. As of today, majority of the activities and tasks related to IoT in the healthcare field revolve around the enhancement of care and comfort along with telemonitoring as one of the main applications in the highly extensive degree of telemedicine [5]. Another zone where different tasks/chores exist includes verifying and complementing resources, making use of IoT and RFID, etc. This is carried out in a fair manner using the gadgets and availed medical service resources [6].

Notwithstanding, these organizations and use cases are only the start and, simultaneously, are a long way from inescapable. Further developed and coordinated methodologies inside the extent of the advanced change of medical care are beginning to be utilized concerning wellbeing information perspectives where IoT assumes an expanding part, as it does in explicit applications, for example, savvy pills, brilliant home consideration frameworks, individual medical care, mechanical technology, and Real-Time Health Systems (RTHS) [5].

12.4 Internet of Things-Based Healthcare Solutions

The dependence of medical care on IoT is expanding continuously to improve admittance to mind, increment the nature of care and in particular decrease the expense of care. In view of a person's remarkable organic, conduct, social and social qualities, the incorporated act of prosperity, medical care and patient help are named as customized medical services. This engages every single individual by following the essential medical services rule of "the correct consideration for the perfect individual at the perfect time," which prompts better results and improvement in fulfilment consequently making medical care financially savvy. Practical help centers around the avoidance, early pathology identification, and homecare rather than the costly clinical one, and checks the general prosperity to foresee needs and guarantee consistence to medical care plans. IoT vows to deal with the personalization of care benefits and can keep up a computerized character for each individual. Distinctive gear is utilized in medical care to impart and to make the pervasive arrangement of framework. The arrangements of

IoT-based customized medical services frameworks are clinical consideration and distant observing [5].

12.4.1 Clinical Care

The frameworks which are not invasive in nature and seem to be observing are made use of for those patients who have been residing in the hospital due to their physiological status and needs. These frameworks and system designs call for a uniform and persistent check and monitoring and make use of the cloud for gathering pertaining data. The collected data is then forwarded to the guardians through private means. Furthermore, it provides a manual and mechanistic progression of data. On this basis, the care being provided is consistently enhanced and improved which also reduces the expense of providing care and also eliminates the need for a guardian [7]. Different models consist of common applications which are not interfunctional, thus shaping the prerequisites in the market for gadgets in the respective field. The connections between the numerous applications in wellbeing observing are:

- The way toward get-together information from sensors (WSNs)
- Backing for standard UIs and presentations.
- Organization network for admittance to infrastructural administrations.
- Being used prerequisites, for example, low force, power, solidness, precision, and dependability.

12.4.2 Distant Checking

The lack of being fully committed toward verifying and cross-checking architectures and frameworks does pose a high risk of many wellbeing chances going undetected and this is a concern to be combatted throughout the globe [8]. Nevertheless, tiny yet impactful remote initiatives interconnected by methods of the IoT ensure that checking is available to patients. Silent information and details related to wellbeing can be proficiently captured with these initiatives or arrangements. A collection of sensors and sophisticated calculations are made use of to decompose the data and later on forward it using remote/local availability. The clinical experts would then be able to make proper wellbeing proposals remotely.

12.5 Machine Learning-Based Healthcare

Medical care is a significant industry which offers esteem-based consideration to a huge number of individuals, while simultaneously turning out to be top income workers for some, countries. Quality, Worth, and Result are three popular expressions that consistently go with medical services and guarantee a ton, and today, medical services trained professionals and partners the world over are searching for creative approaches to convey on this guarantee. Innovation empowered brilliant medical care is not, at this point a trip of extravagant, as Web associated clinical gadgets are holding the wellbeing framework as we probably are aware it together from self-destructing under the populace trouble.

From assuming a basic function in patient consideration, charging, and clinical records, today innovation is permitting medical care experts create substitute staffing models, IP capitalization, give savvy medical care, and lessening regulatory and gracefully costs. AI in medical care is one such territory which is seeing slow acknowledgment in the medical care industry. Google as of late built up an AI calculation to recognize carcinogenic tumors in mammograms, and specialists in Stanford College are utilizing profound figuring out how to distinguish skin disease. ML is now helping out in assorted circumstances in medical care. ML in medical services assists with breaking down huge number of various information focuses and recommend results, give ideal danger scores, exact asset allotment, and has numerous different applications. In this article, we will examine a portion of the top uses of AI in medical care, and how they remain to change the manner in which we envision the medical care industry in 2018 and beyond. We accept that medical care suppliers need to quit considering AI as an idea from the future and rather grasp this present reality devices today are making it accessible to us! Throughout the long term, we have helped worldwide medical services a customer influence the most recent in innovation to help patients and partners the same. With regards to AI, we discover explicit use cases in which ML-based applications can offer something of unmistakable benefit to your medical care activities, and afterward help build up a bit by bit cycle to fuse the equivalent inside your cycles.

12.5.1 Future Model of Healthcare-based IoT and Machine Learning

As discussed in the above sections, the medical arena is likely to possess advancement with the integration of IoT- and ML-based components in the future. Lately, medical care is progressively utilizing data advances for conveying savvy frameworks pointed toward accelerating wellbeing diagnostics and therapy. Such frameworks offer keen types of assistance for observing wellness and clinical robotization in various settings and conditions (clinics, workplaces, home, on-the-go, etc.), hence permitting a considerable decrease of doctor visit costs and an overall upgrade of patient consideration quality.

In this specific circumstance, the wide dissemination of incredible installed equipment along with the advancement of brilliant clinical sensors and gadgets for universal medical care has made the Internet of Medical Things (IoMT) definitely change the manner in which medical care is moved toward around the world, so the quantity of medical care gadgets utilizing IoT and wearable innovations is required to arrive at 162 million before the finish of 2020. Information caught by wearable, ingestible, and inserted sensors, versatility designs, and gadget utilization designs permit to follow client propensities and can be viably gathered and handled to uncover basic conditions by utilizing best in class Artificial intelligence (artificial intelligence) and Machine/Deep learning (ML/DL)-based methodologies. Conventional cloud-based designs for Huge Information investigation can give great execution and dependability when supporting non-security and inertness basic IoT applications. However, when the end client is a patient with basic and time-touchy necessities, a more serious level of power and availability is needed, since the event of detachment from the center organization or transfer speed/inactivity varieties could have a drastically negative effect and furthermore lead to lethal results in crisis circumstances.

Developing interest toward designs that understand the collaboration of Cloud, Mist, and Edge processing is recently arising. The principle objective is to abuse the maximum capacity of edge hubs and low-level haze hubs to deal with practical undertakings as well as information handling, examination, relationship, and induction. Such methodologies speak to a promising answer toward the execution of dependable circulated medical care applications and administrations, since a savvy planning of computational and asset the board errands over the hubs demonstrates to meet the rigid prerequisites of IoMT situation. In this specific situation, we help to the spread of "Edge/Mist Wellbeing" arrangements that utilize appropriate registering ideal models to disseminate wellbeing sensors information preparing and capacity among various hubs, situated at various degrees of vicinity to clients, as follows:

- Edge registering happens straightforwardly on the gadgets to which the sensors are appended or a door gadget that is truly near the sensors: instances of edge hubs are wearable gadgets, for example, cell phones, smartwatches or versatile "impromptu" inserted frameworks like single-board PCs, microcontrollers;

- Mist processing hubs act at a neighborhood level, so they can incorporate greater and all the more remarkable gadgets, for example, laptops, nearby workers and passages that might be actually more removed from the sensors and actuators.

The two ideal models (progressively frequently actualized together) influence the nearness to the client to furnish area mindful wellbeing administrations with diminished inertness and high accessibility. A few techniques depending on progressive processing methodologies have been proposed to allot and disseminate the deduction errands of artificial intelligence and ML strategies between the cloud, the haze and the edge levels (or mist/edge peers), attempting to push the (restricted) computational limits of edge gadgets to their top. A change from the mobile cloud computing model (MCC), described by high information transmission expenses and restricted inclusion toward a multi access edge computing model (MEC) with low-inertness and solid edge ML models is then dynamically occurring in the keen medical services space.

12.6 Wearable System for Smart Healthcare

Wearable Healthcare Device (WHDs) are an arising innovation that empowers ceaseless wandering and observing of human crucial signs during life, day by day (during work, at home, during sport exercises, and so forth), or in a clinical climate, with the benefit of limiting inconvenience and obstruction with typical human exercises. WHDs are essential for individual wellbeing frameworks, an idea presented in the last part of the 1990s, to put the individual resident in the focal point of the medical care conveyance measure, dealing with its own wellbeing and connecting with care suppliers—an idea that is ordinarily alluded to as "tolerant strengthening." The point was to raise individual's interest about their wellbeing status, improving the nature of care, and utilizing the new innovation capacities. These gadgets make a collaboration between various science spaces, for example, biomedical advancements, miniature and nanotechnologies, materials designing, electronic designing, and data and correspondence innovations.

The utilization of WHDs permits the wandering securing of fundamental signs and wellbeing status checking over broadened periods (days/weeks) and outside clinical conditions. This component permits obtaining fundamental information during various day by day exercises, guaranteeing a superior help in clinical determination or potentially helping in a superior and quicker recuperating from a clinical mediation or body injury. WHDs are likewise extremely valuable in game exercises/wellness to screen competitor's exhibition or even in specialists on call or military work force to assess and screen their body reaction in various perilous circumstances and to more readily deal with their work- and word-related wellbeing. These gadgets can be for both clinical and additionally exercises/wellness/wellbeing purposes, continually focusing on the human body observing. Considering, the best wording is "wellbeing," prompting WHDs. WHDs group can be more explicit alluding to which zones they are applied to. Autonomously of WHDs reason, there are four primary necessities on their plan: low force utilization, unwavering quality and security, solace, and ergonomics.

As of late, we help to a truly developing dissemination of brilliant clinical sensors and IoT gadgets that are intensely changing the manner in which medical services are moved around the world. In this specific situation, a blend of Cloud and IoT structures is frequently misused to make brilliant medical services frameworks fit for supporting close real-time applications when handling and performing AI on the enormous measure of information created by wearable sensor organizations. Anyway, the reaction time and the accessibility of cloud-based frameworks, along with security protection, actually speak to a basic issue that forestalls IoMT gadgets and designs from being a solid and powerful answer for the point. Of late, there is a developing interest toward designs and approaches that adventure Edge and Haze processing as a response to repay the shortcomings of the cloud. In this paper, we propose a short audit about the overall utilization of IoT arrangements in medical care, beginning from early wellbeing checking arrangements from wearable sensors up to a conversation about the most recent patterns in haze/edge registering for shrewd wellbeing.

12.7 Communication Standards

The consideration of patients now definitely appears to include a wide range of people, all expecting to share quiet data and talk about their administration. As an outcome, there is expanding interest in, and utilization of, data and correspondence innovations to help wellbeing administrations. Surely, in the event that data is the soul of medical services, at that point correspondence frameworks are the heart that siphons it. Yet, while there is huge conversation of, and interest in data advances, correspondence frameworks get substantially less consideration. While there is some critically progressed research in exceptionally explicit regions like telemedicine, the clinical selection of significantly less complex administrations like voice message or electronic mail is as yet no ordinary in numerous wellbeing administrations. A lot of this would change in the event that it were all the more broadly understood that the greatest data archive in medical care sits in the top of individuals working inside it, and the greatest data network is the intricate trap of discussions that connect the activities of these people [9]. There are tremendous holes in our wide comprehension of the part of correspondence administrations in medical services conveyance. Lab medication is maybe significantly more ineffectively concentrated than numerous different zones, for example, the interface between essential consideration and clinic administrations. However, clinical research centers from various perspectives are message-preparing undertakings, accepting messages containing data demands, and producing results that are sent as messages back to clinical administrations [9]. While there is a lot of current spotlight on improving lab pivot times and inside efficiencies, little is truly thought about the more extensive correspondence measures inside the medical care framework, of which clinical research centers are nevertheless one connection in the chain [10]. However, without this more extensive view, there is always present a danger that neighborhood frameworks inside research facilities are enhanced and overdesigned, yet that the worldwide execution of wellbeing administrations remains moderately unaltered.

Given this absence of explicit data about research facility correspondence benefits, this paper will venture back, and by and large survey the segments of a correspondence framework, including the fundamental ideas of a correspondence channel, administration, gadget, and collaboration mode. The survey will at that point attempt to sum up some of what is thought about explicit correspondence issues that emerge across wellbeing administrations in the principle, including the network and clinic administration conveyance [9].

12.8 Challenges in Healthcare Adoption with IoT and Machine Learning

Despite the fact that IoT in medical services gives numerous incredible advantages, there are likewise a few moves that should be unraveled. The Internet of Things Medical services arrangements can't be considered for execution without recognizing these difficulties.

- Enormous contributions of produced information. Having a great many gadgets in a solitary medical services office and 1,000 additionally sending data from far off areas—all progressively—will produce tremendous measures of information. The information produced from IoT in medical care will probably cause stockpiling necessities to develop a lot higher, from Terabytes to Petabytes. Whenever utilized appropriately, artificial intelligence driven calculations and cloud can help figure out and sort out this information, yet this methodology needs an ideal opportunity to develop. Along these lines, making a huge scope IoT medical care arrangement will take a great deal of time and exertion.

- IoT gadgets will build the assault surface. IoT medical services carry various advantages to the business; however they additionally make various weak security spots. Programmers could sign into clinical gadgets associated with the Web and take the data—or even change it. They can likewise make a stride further and hack a whole clinic organization, tainting the IoT gadgets with the notorious Ransomware infection. That implies the programmers will hold patients and their pulse screens, circulatory strain perusers, and cerebrum scanners as prisoners.
- Existing programming framework is outdated. IT frameworks in numerous medical clinics are outdated. They won't take into account appropriate joining of IoT gadgets. Thusly, medical care offices should patch up their IT cycles and utilize new, more current programming. They will likewise have to exploit virtualization (innovations like SDN and NFV), and super quick remote and versatile organizations like Progressed LTE or 5G.

12.9 Improving Adoption of Healthcare System with IoT and Machine Learning

A portion of the manners by which IoT and enormous information can together assist to improve the appropriation of keen wellbeing which will bring about improved medical services conveyance and access.

12.9.1 Proof-based Consideration

The outstanding expansion in the volume of medical services information created by IoT gadgets makes information handling testing. Enormous information can give proof-based consideration by conglomerating informational indexes from assorted sources. Examination of information can give helpful bits of knowledge into recognizing irregularities and giving suitable medicines to patients. Savvy investigation utilizing new strategies can give generous monetary reserve funds on the request for a few hundred billion dollars, which adds up to around 8 percent of the public wellbeing costs (Olaronke and Oluwaseun, 2016). The investigation of wellbeing related data with productive techniques advances early recognizable proof of sickness designs, which extends general wellbeing reconnaissance. This guarantees that proper and opportune choices on the treatment of a specific infection are taken accordingly decreasing patient mortality. Huge information upgrades the sort of care patients get as treatment choices depend on information accumulated from examining enormous informational collections.

12.9.2 Self-learning and Personal Growth

The various sensors of IoT envision assortment and collection of information/data. However, IoT on its own isn't capable of producing recovery of medicines. Accurate and convenient therapies can be developed to be reliable on swift patient test, and the enhancement of recovery techniques comparing to the assessments from the clinic. Different parameters need to be taken into account in order to provide an accurate treatment. PC gadgets and gizmos mostly rely on the data that has been collected through the sensors and previous analyses, whereas self-learning techniques can casually explore and come up with novel treatment alternatives. Geography-related and cosmology-based heuristic computations can boost the process of finding ideal solutions for a massive scope medical service routine (Yuehong et al., 2016) [11]. Different circulated processing stages are being utilized today for large information examination. These stages incorporate Apache Samza, Apache Sparkle, Hadoop MapReduce, Apache Tempest, and Flink. Hadoop MapReduce and Apache Sparkle are the most generally utilized stages for gigantic information stockpiling and examination (Praveena and Bharathi, 2017). Hadoop is a simple to utilize open-source instrument for dealing with enormous information applications [12]. The Hadoop MapReduce system gives a significant dispersed figuring stage that is equipped for putting away and handling a lot of unstructured informational indexes (Khan and Iqbal, 2017). MapReduce (Merla and Liang, 2017) is a programming climate that licenses equal and conveyed preparing on tremendous measures of information on enormous bunches of equipment. Hive (Garg, 2015) is the organized inquiry language (SQL)–like extensions that license unsurprising business applications to run SQL inquiries against a Hadoop bunch. PIG (Jain and Mayrya, 2017) is an apparatus that makes Hadoop more usable by making MapReduce questions less difficult to execute. Wibidata (Moorthy et al., 2014) is an apparatus that coordinates Hadoop with Web investigation to upgrade information use by sites. It is a stage that naturally maps client's questions to Hadoop occupations. Rapidminer (Dwivedi et al., 2016) gives an incorporated stage to investigation (both business and prescient), mining of information and AI [12].

12.9.3 Normalization

Various groups like IEEE, IETF, etc. have contributed to the normalization and alignment of innovations in the field of IoT. The normalizations were mainly impacted by the feedbacks of the European Telecommunication Standards Institute (ETSI) and Internet Engineering Task Force (IETF) working associations [12]. All nascent and arising thoughts must be consolidated in order to curate a worldwide arrangement that would help build normalizations for the forthcoming Web. In light of the outcomes gave by the CERP-IoT venture (IERC, 2016), future Web is an expansion of the current one by incorporating general things into more extensive organizations. The normalization will empower the advancement of IoT-based medical care frameworks.

12.9.4 Protection and Security

In the current decade, with rapid advancements in science and technology, protection and privacy poses a grave concern. IoT-based frameworks are helpful as long as its clients stay safe. In IoT frameworks, a wide range of information assortment and mining are performed over the Web. Hence, individual information can be gotten to at different stages (during assortment, transmission, etc). Patients' security should be contemplated by forestalling any type of following or unlawful recognizable proof [12]. The higher the degree of self-sufficiency and knowledge of the IoT gadgets, the harder the security of personalities and protection becomes. IoT-based applications are likewise weak on account of remote correspondence which makes snooping simpler. Furthermore, IoT gadgets for the most part have low energy and low registering power which makes it harder to execute complex calculations to ensure security. As large information turns out to be more universal in the medical care framework, greater security difficulties will arise. Thorough examination is expected to guarantee protection, trust, and security all through the medical services climate [11].

12.9.5 Intelligent Announcing and Representation

Huge information applications need to recognize examination and reports (Suyts et al., 2017). Huge information applications would not succeed if information are essentially composed to reports. Applications need to get important experiences from a majority of information and just notice explicit features (intelliPaat, 2019). It is likewise important to prepare calculations to produce exact bits of knowledge dependent on accessible information without which the validity of the report comes into question. Reports can be made engaging and valuable by including charts and factual data. Applications ought to likewise zero in on creating perceptions that would make it simple to get experiences from a report and permit simple ID of patterns and difficulties in a medical services section.

12.10 Proposed Solution Based on IOT and Machine Learning for Smart Healthcare Systems

This section elucidates the solution and framework we wish to propose in the field of healthcare. IoT is a structure that utilizes advancements like sensors, network correspondence, man-made brainpower, and big data to give genuine arrangements. These arrangements and frameworks are intended for ideal control and execution.

- IoT is an occurrence Innovation given the headways in united advances like Sensor, Correspondence and Processing. With these headways, any leaf hub gadget of today is fit for "detecting" its environmental factors, can perform calculation and is addressable by an organization address over a remote organization. This empowers answers for be built up that can plan "reality" elements to a comparing virtual item. These virtual articles can speak with each utilizing accessible correspondence advances and keep the "reality" substance educated about the condition of "things". A control instrument between the "reality" elements and the virtual articles is likewise included as a component of this structure/arrangement.
- Cloud computing involves themes identified with giving figuring administrations and utilities like workers, stockpiling, information bases, organizing, programming, investigation, and so forth over the Internet. Contingent upon the prerequisites of the end client, different administrations can be given from a far-off area.

196 Research Issues and Future Research Directions

- Edge computing is a subset of cloud computing where huge number of these administrations is given from an area that is geologically conclusive to the end client and can in this manner effectively eliminate network inactivity.

A run of the mill-IoT model contains an end hub gadget that can speak with a back-end calculation/server farm over a correspondence medium (By and large Remote). The correspondence channel generally utilizes IoT conventions like MQTT/CoAP. Information and Control Messages can be flawlessly traded over the IoT endpoint and a Server farm Worker. The Endpoint gadget can give encompassing condition data to a Server farm utilizing different Sensors relying upon the space of intrigue and get back data/guidelines from backend to perform activities. The Back-End Workers are commonly groundbreaking figuring assets and can utilize concentrated calculations to deal with the information accumulated from the End Point gadgets. The principle challenges around IoT arrangements are:

- The measures of information created by the sensors are tremendous. Extraction of applicable data from the caught information is a test. This exertion requires advancement of a calculation that can extricate anomalies in caught information for body sensor organizations. There have significant exploration scopes in field of AI and testing calculations.
- Given the way that calculation serious tasks are pushed to back end, streamlining of Continuous Reaction is a zone of progress. Upgrading the measure of information move is a region of interest.
- Decentralization of calculation. With an ever increasing number of gadgets being IoT fit, calculation at one point will make bottleneck in organization assets. The calculation should be circulated and Errand Level Parallelism should be accomplished. Calculation and asset appropriation calculations are zones of significant exploration interest in this field.
- Security of the IoT gadgets.
- Force Utilization at End Point Gadgets.

The evolving/handling of IoT gadgets/devices is not be a simple task. The maintenance of IoT devices comes with a cost. This issue is commonly settled by offloading undertakings to a back-end worker and sparing battery power that would have been generally needed for in-house processing. This gave a significant driving force to investigate in the areas of decentralization of calculation.

Edge figuring encourages network reaction times, helps decentralization, and can likewise address security concerns. This is on the grounds that a huge piece of the basic calculation would now be able to be performed at the "edge hubs" which would interface with cloud intermittently. This gives the offices of cloud computing sans its weaknesses. Combined with AI and Huge information apparatuses, high adequacy ongoing arrangements can be created.

In this examination, we are investigating the conceivable outcomes of joining of cloud/edge figuring and AI standards into an Appropriated processing-based IoT Structure. The objective is to ready to separate applicable data of premium among the immense information that is ordinarily produced by the front-end Sensor structures in IoT gadgets. Some insight can be remembered for the front-end module itself to empower the front-finish to take a choice on information need. Direction with respect to how to accomplish this can be given by a backend IoT worker. It is recommended that the backend worker has AI-based usage to have the option to naturally learn information marks of revenue dependent on the information it has just gotten.

As a utilization instance of the abovementioned, we intend to apply the above ideas to Clinical applications. There has been a plenty of clinical sensors right now accessible like:

- SPO2 Sensor
- ECG Sensor
- Wind stream Sensor
- Temperature Sensor
- Sphygmomanometer
- Body Position Sensor
- Galvanic Skin Reaction Sensor
- Glucometer
- EMG Sensor

Assurance of if an information is basic should be taken by the leaf gadget with direction from the backend cloud/edge worker. The cloud/edge worker should have the option to take in data from the current leaf hub just as different hubs it serves and give rules to the endpoint gadget with respect to need of choices. The choice would be founded on conventional information accessible according to clinical records just as customized information produced.

Genuine utilizations of the above methodology would be:

- Decide abrupt circulatory strain variances. Examine these vacillations and check whether these are abnormalities or not and ready crisis

benefits appropriately. Such information is basic and needs continuous consideration and, thusly should be organized over others. Some information examples may be typical for certain patients yet for nobody else.
- Decide body pose developments of the patient and check when the patient is requiring consideration for development. This can be utilized for helping patients and the arrangements required for them [20].
- Decide epilepsy seizure dependent on the investigation from information accessible from electroencephalography [29]. Work on calculations with the goal that the seizure can be identified at a registering asset close to the sensor.
- Work on a calculation/arrangement that will organize transmission and handling of basic information over non-basic information.
- Guide a rescue vehicle to suitable wellbeing community that is the nearest, having significant offices dependent on patient's wellbeing exam information gathered by Body Sensor Organizations.

In last, all readers are requested to read our papers [13–21] to know more about Internet of Things, Internet of Things based Healthcare, Machine Learning role in Healthcare, Deep Learning, etc.

12.11 Conclusion

Late progressions in "network" and "detecting" innovations in endpoint gadgets combined with cloud computing has acquired examination interests in IoT-based answers for Medical care, helped living, farming, and so forth. As a piece of this investigation, we will view a portion of Anxious Processing in IoT, while zeroing in on medical services advancements. We saw that numerous medical services arrangements require continuous dynamic capacities. Such an answer doesn't lean toward cloud computing due to the organization deferrals and latencies that are related with it. A working model of such an answer has additionally been proposed. Such an answer can control the endpoint IoT gadget utilizing IoT conventions like MQTT and simultaneously gather data from cloud and play out the offloaded activities. As a piece of our further examination, we intend to consider various IoT edge workers, their associations with one another and the End Point gadgets also.

References

[1] https://myassignmenthelp.com/free-samples/swe30011-iot-programming/internet-of-things-for-smart-cities.html

[2] Malasinghe, Lakmini P., Ramzan, Naeem, Dahal, Keshav, Remote patient monitoring: a comprehensive study, Journal of Ambient Intelligence and Humanized Computing, 2019.

[3] Damian Dziak, Bartosz Jachimczyk, and Wlodek J. Kulesza, IoT-Based Information System for Healthcare Application: Design Methodology Approach, MDPI, 2017.

[4] B. Bansal and M. Sharma, "Client-Side Verification Framework for Offline Architecture of IoT," 2019 3rd International conference on Electronics, Communication and Aerospace Technology (ICECA), 2019.

[5] Pethuru Raj, Jyotir Moy Chatterjee, Abhishek Kumar, B. Balamurugan, Internet of Things Use Cases for the Healthcare Industry.

[6] Robert S. Kaplan and Michael E. Porter, the Cost Crisis in Health Care, https://hbr.org/2011/09/how-to-solve-the-cost-crisis-in-health-care

[7] Saavitha C, Dr. Sharada Prasad, Survey on IOT based Architectures in Healthcare Applications, International Journal of Combined Research & Development (IJCRD), Volume: 5; Issue: 7; July -2016.

[8] Purcell, Rosemary, Gwyther, Kate, Rice, Simon M, Mental Health In Elite Athletes: Increased Awareness Requires An Early Intervention Framework to Respond to Athlete Needs, Sports Medicine – Open, 2019.

[9] Levit L, Balogh E, Nass S, Patient-Centered Communication and Shared Decision Making, Board on Health Care Services, 2013.

[10] Dash, S., Shakyawar, S.K., Sharma, M. et al. Big data in healthcare: management, analysis and future prospects. J Big Data 6, 54, 2019.

[11] M.N. Mohammed1, S.F. Desyansah, S. Al-Zubaidi3, E. Yusuf, An internet of things-based smart homes and healthcare monitoring and management system: Review, Journal of Physics: Conference Series, 2020.

[12] Donal Tobin, https://www.xplenty.com/blog/apache-spark-vs-hadoop-mapreduce/

[13] B. Gudeti, S. Mishra, S. Malik, T.F. Fernandez, A.K. Tyagi and S. Kumari, "A Novel Approach to Predict Chronic Kidney Disease using Machine Learning Algorithms," 2020 4th International Conference on Electronics, Communication and Aerospace Technology (ICECA), Coimbatore, 2020, pp. 1630–1635, doi: 10.1109/ICECA49313.2020.9297392.

[14] Nair M.M., Kumari S., Tyagi A.K., Sravanthi K. (2021) Deep Learning for Medical Image Recognition: Open Issues and a Way to Forward. In: Goyal D., Gupta A.K., Piuri V., Ganzha M., Paprzycki M. (eds) Proceedings of the Second International Conference on Information Management and Machine Intelligence. Lecture Notes in Networks and Systems, vol 166. Springer, Singapore. https://doi.org/10.1007/978-981-15-9689-6_38

[15] Akshara Pramod, Harsh Sankar Naicker, Amit Kumar Tyagi, "Machine Learning and Deep Learning: Open Issues and Future Research Directions for Next Ten Years", Book: Computational Analysis and Understanding of Deep Learning for Medical Care: Principles, Methods, and Applications, 2020, Wiley Scrivener, 2020.

[16] Tyagi, Amit Kumar and G, Rekha, Machine Learning with Big Data (March 20, 2019). Proceedings of International Conference on Sustainable Computing in Science, Technology and Management (SUSCOM), Amity University Rajasthan, Jaipur - India, February 26–28, 2019.

[17] Amit Kumar Tyagi, Poonam Chahal, "Artificial Intelligence and Machine Learning Algorithms", Book: Challenges and Applications for Implementing Machine Learning in Computer Vision, IGI Global, 2020. DOI: 10.4018/978-1-7998-0182-5.ch008

[18] Kumari S., Vani V., Malik S., Tyagi A.K., Reddy S. (2021) Analysis of Text Mining Tools in Disease Prediction. In: Abraham A., Hanne T., Castillo O., Gandhi N., Nogueira Rios T., Hong TP. (eds) Hybrid Intelligent Systems. HIS 2020. Advances in Intelligent Systems and Computing, vol. 1375. Springer, Cham. https://doi.org/10.1007/978-3-030-73050-5_55

[19] Amit Kumar Tyagi, G. Rekha, "Challenges of Applying Deep Learning in Real-World Applications", Book: Challenges and Applications for Implementing Machine Learning in Computer Vision, IGI Global 2020, p. 92–118. DOI: 10.4018/978-1-7998-0182-5.ch004

[20] Varsha R., Nair S.M., Tyagi A.K., Aswathy S.U., RadhaKrishnan R. (2021) The Future with Advanced Analytics: A Sequential Analysis of the Disruptive Technology's Scope. In: Abraham A., Hanne T., Castillo O., Gandhi N., Nogueira Rios T., Hong TP. (eds) Hybrid Intelligent Systems. HIS 2020. Advances in Intelligent Systems and Computing, vol. 1375. Springer, Cham. https://doi.org/10.1007/978-3-030-73050-5_56

[21] Tyagi, Amit Kumar; Nair, Meghna Manoj; Niladhuri, Sreenath; Abraham, Ajith, "Security, Privacy Research issues in Various Computing Platforms: A Survey and the Road Ahead", Journal of Information Assurance & Security. 2020, Vol. 15 Issue 1, p. 1–16. 16 p.

13

A Novel Adaptive Authentication Scheme for Securing Medical Information Stored in Clouds

N. Moganarangan[1], N. Palanivel[2], and S. Balaji[3]

[1]Associate Professor Department of Computer Science and Engineering
Rajiv Gandhi College of Engineering and Technology, Pondicherry, India
[2]Associate Professor Department of Computer Science and Engineering
Manakula Vinayagar Institute of Technology, Puducherry, India
[3]Associate Professor Dept. of Information Technology Sri Manakula
Vinayagar Engineering College Puducherry. India
E-mail: rengannsj77@gmail.com; npalani76@gmail.com;
balajisankaralingam@gmail.com

Abstract

Medical cloud systems are common in the recent years due to distributed and pervasive access to information without requiring additional infrastructure units. As the environment is distributed, access and security are the fundamental requirements for storing and retrieving sensitive medical information. In this chapter, adaptive authentication scheme for securing the stored medical information is presented. This authentication scheme is designed on the basis of access control and requesting time instances, to improve the authentication and information retrieval. This scheme supports both store and update processes in the medical cloud from which the access authentication is deliberated to the end-users. Based on the above mentioned process, the session access key for the end-users are updated to prevent anonymous access and medical information retrieval. The proposed scheme is verified through simulations and the metrics process delay, integrity check bytes, and overhead are assessed using a comparative analysis.

Keywords: Access control, authentication, cloud, medical data.

13.1 Introduction

In the recent years, diagnosis centers and hospitals rely on information communication technologies (ICTs) for handling medical data for ease of computation and access. The raw medical data/information accumulated from the patients/end-users are converted into electronic format called electronic health records (EHRs) [1]. These records are stored in dedicated medical cloud from where the access to the information is distributed to the end-users. The need for pervasive computing, access and storage is increasing due to the on-demand requirements of the end-users and lack of storage. With the evolution of cloud computing and its associated features such as storage, computation as service, and electronic medical records are stored in such environments [2]. The computation, storage, and access in these environments are mutually shared by the medical centers with the end-users to provide timely update regarding their health and physiological observations. Based on this conceptual requirement, healthcare cloud and medical information systems are designed. These systems and storage clouds provide both computational and communication-oriented services for the end-users through different service providers and application support [2, 3]. For this purpose, it exploits heterogeneous communication technologies and integrated environment with multiple applications for an on-demand access [4].

The cloud environment is distributed and hence it consists of multiple constraints including security and privacy. Medical information is sensitive and it has to be prevented from anonymous access and attacks to retain its trust ability [4, 5]. Public or dedicated medical cloud storage requires precise security mechanisms to protect the EHR and to ensure the information is privacy protected. This distributed environment ensures heterogeneous access to the information stored for a different density of users [6]. Therefore, centralized and independent security measures are required for this storage and processing system to ensure privacy for the connected users. Besides, access control is another requirement to restrict user access and information retrieval from anonymous sources [7]. User authorization, access control and information authentication is required for providing privacy and security of the stored medical information in the cloud. Administering security through authentication protocols, third party servers, peer-to-peer authorization, etc., are widely adopted in medical clouds. However, the need for security is still demanding with the increase in volumes of medical information and attack types [5, 7].

Tang et al. [8] introduced the concept of blockchain for providing authentication for EHRs. Signatures generated using the identity of the communicating users in this method helps to mitigate the impact of collusion attacks.

Light-weight privacy preserving medical services access (LPP-MSA) is designed by Liu et al. [9] for providing security for healthcare cloud resources. With the help of online and offline signing and third party verification, this method achieves anonymity and resistance to collusion attacks.

In [10], another identity-based privacy preserving method is introduced for Internet of Things (IoT)-based health storage systems. Data authentication is preceded using edge servers and the integrity verification provided by these servers reduces the computation and management cost.

Zhou et al. [11] introduced access flexible security mechanism for medical data for ensuring the privacy of stored information. This method also adopts online/offline authentication mechanism for improving the security levels and processing speed of the information across the distributed storage.

Mutual authentication is employed by Li et al. [12] for securing the cloud-dependent telecare medical information systems. This authentication is administered between the medical information systems and the storage cloud for granting privacy for the stored information. This authentication mitigates unlink ability issues in data access by retaining the anonymity of the users.

The authors in [13] discussed the chances of data de-duplication for preventing data wastage in medical clouds. The de-duplication problem is addressed using convergence-based encryption. Along with this encryption method, bloom filters are employed for improving the efficiency of medical searches using keywords.

Considering the significance of medical data sharing, Masud and Hossain [14] designed a secure-data exchange protocol for health care systems. This protocol facilitates data storage and management across different cloud environments using two-phase security. In this protocol, secret sessions are generated for exchanging information using elliptic curve cryptography. This security method is resilient against man-in-middle and replication kind of attacks.

Kaur et al. [15] analyzed the feasibility and future directions for employing blockchain medical clouds. In order to provide security and to retain the freshness of the heterogeneous medical data stored in the cloud, blockchain-based solutions are discussed by the authors. The authors highlight the accuracy, cost effectiveness, and security features of the medical cloud data on using a blockchain.

13.2 Adaptive Authentication Scheme

The design goal of adaptive authentication scheme is to provide privacy and security for medical data shared and stored in medical cloud. Different from the conventional methods of two-way or third party-based authentication, sequence-based authentication is adapted in this scheme. This scheme endorses three components namely diagnosis center, medical cloud, and the end-user. The access and authentication is shared between the diagnosis center and the end-user in the cloud. Access and update sequence are the key factors determining the security features of the medical data shared in the cloud. The architecture of the proposed scheme is represented in Figure 13.1.

As represented in Figure 13.1, the architecture is modeled with the communication/interaction sequence between the three components described. In the following Table (Table 13.1), the functions and the description of the components are detailed.

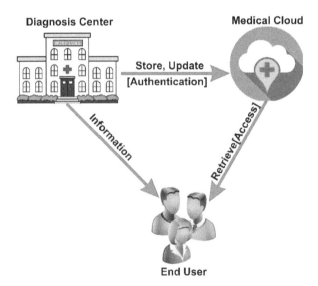

Figure 13.1 Architecture of the proposed scheme.

Table 13.1 Components and their functions

Component	Functions	Security Measure	Description
Diagnosis Center	Aggregates information from the end-user. Stores/ updates information in the cloud	Authentication, Access Control	Provides authentication for the stored/ aggregated medical information
Medical Cloud	Allocation of storage space for the aggregation medical data	Decryption Correlation	Ensures access to the end-users as identified by the diagnosis center
End-user	Shares medical data through physiological investigations at diagnosis center. Retrieve his/ her medical information form the cloud	Access Control, Authentication	Decrypting the stored medical information through authenticated access control

The proposed authentication scheme follows elliptic curve cryptography-based medical data security along with access control. Unlike the traditional ECC, the random integer (I) of the algorithm is generated from the range of access and update sequence. The access and update sequence from the diagnosis center to the medical cloud is considered for authentication and information retrieval access. The diagnosis center stores the medical information as digital record in the cloud. The record is accessed through the device used by the end user. Let m denote the medical data/information, aggregated from the patient (end user). This information contains the physical logical observations of the end-user recorded in the diagnosis center. Each m is classified using unique patient ID. Access to the m in the cloud is provided to the end-user by mapping the ID and the access list. The access list contains the set of authenticated users and described by the diagnosis center. A user registered with the diagnosis center is allocated with this ID and is said to be authenticated.

13.3 Information Storage/Update

The diagnosis center establishes communication sessions with the medical cloud. These sessions are secured through session keys that are generated using time synchronization factor (t_s). The cloud service provider makes use of both public (P_c) and private (Q_c) keys for t_s with the public and private keys of the information server (p_s) and (Q_s). This information server (s) is retained by the diagnosis center. There are two session keys (S_k) generated depending on t_s (i.e.)

$$S_k = \begin{cases} \prod_{i=1}^{n} \frac{i}{t_{s_i}} |I|, & \forall\, t_s = 1 \\ \prod_{i=1}^{n-t_s} \frac{i}{1-t_{s_i}} |I|, & \forall\, 0 < t_s < 1 \end{cases} \quad (13.1)$$

In Equation (13.1), n represents the number of m that is to be stored. Similarly, if the request and acceptance time of diagnosis center and medical cloud are same, then $t_s = \frac{t_a}{t_r}$, where t_a and t_r are the time for acceptance and request corresponding. In the first request process, $t_r > t_a$ and therefore it follows the S_k as per $0 < t_s < 1$ condition. The variable I in a storage process is estimated as

$$I = \begin{cases} \frac{r_1}{P_c} - \frac{t_s}{n}, & if\ it\ is\ a\ storage \\ \frac{r_2}{P_c} - \frac{t_s}{n-t_s}, & if\ it\ is\ an\ update \end{cases} \quad (13.2)$$

Where r_1 and r_2 are the count of storage sequence and update sequence. This sequence is confined to the Q_s of the diagnosis center. Therefore, the authentication process is defined as

$$\left. \begin{array}{l} m_s \simeq [hash\,(r_1, m) + I\,Q_c]\,|r_1 - n| \\ m_u \simeq [hash\,(r_2, m) + IQ_c]\,|r_2 - n - t_s] \end{array} \right\} \quad (13.3)$$

Here, m_s and m_u are the encoded message new storage and m_u is the encoded message for update respectively. In both the cases, the condition of $[m_s Q_s + r_1 P_c] = [m_s Q_c + r_1 P_s]$ or $[m_u Q_s + r_2 P_c - t_s] = [m_u Q_c + r_2 P_s - t_s]$ is satisfied, then the message m is said to be encrypted (authenticated). This information is stored in the cloud along with the medical data access list. In the access list, the ID and access time of the user are accounted. However, the user is provided with a single decryption function for both the m_s and m_u visualization. Therefore, the security session for data retrieval is established using the access sequence and ID of the user. Similarly, if the information stored in the cloud exceeds the access time interval or been accessed with a miscommunicated ID, then this session is discarded. In Figure 13.2(a)

13.3 Information Storage/Update

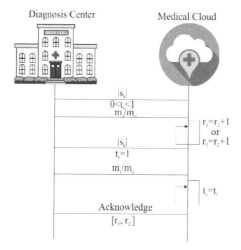

Figure 13.2(a) Request process.

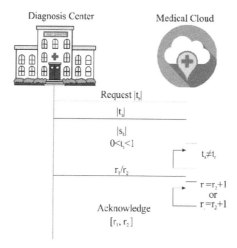

Figure 13.2(b) Message/Update process.

and 13.2(b) the process of request and message storage/update between the diagnosis center and medical cloud is illustrated.

In Figure 13.2(a) and Figure 13.2(b) the process illustrates that if $t_s \neq t_r$, then S_k follows $r_1 = 2$ or $r_2 = 2$ sequence (or) $r_1 + r_2 = 2$ sequence such that t_s is shared in common by the S_k between the diagnosis center and medical cloud. This change in sequence pursues an incremented chain of r_1 or r_2 depending upon m_s or m_u transmission. The storage/update of the medical record/data provides access control based on the r_1 and r_2 that is valid until the user access. Let t_p denote the process time (either for update/storage) under multiple instances such that the querying time (t_q) is greater than t_p. The querying time is generated from the user device for accessing the stored or updated m in the medical cloud. Now, there are two cases for providing authentication access to the end-users such that either $t_p > t_q$ (or) $t_p \leq t_q$. These two cases are analyzed in a differential manner to provide authentication in a better manner. The case of $t_p > t_q$ is optimal provided the access is given to the end-user with the updated S_k and I. Instead, the case of $t_p \leq t_q$ is not optimal as the access requires updated and uninterrupted flow of information view. Therefore, the S_k for $t_p < t_q$ and $t_p = t_q$ is given as

$$S_{ku} = \begin{cases} \prod_{i=1}^{(t_q-t_p)} \frac{i}{t_{p_i}} - q_i \times t_{q_i}, & if\ t_p < t_q \\ \prod_{i=1}^{t_p} (i \times t_{q_i}) - \frac{q_i}{t_{q_i} - t_{p_i}}, & if\ t_p = t_q \end{cases} \qquad (13.4)$$

Where, S_{ku} is the session key generated by the user. For this case, the end-user relies on its private and public (Q_u, P_u) key generated at the time t_q. The private key of the user does not change whereas the public key for accessing the medical information changes with t_q. The update for S_{ku} relies on the P_u and Q_u of the end-user. The change in session key between the medical cloud and the end-user is updated if $t_p < t_q$, ensuring non-redundant and delay less access to the m.

13.4 Integrity Check

The integrity verification ensures date freshness and precise information availability in the medical cloud. This integrity check provides minimum overhead in time-based validation. The synchronization factor is the starting time instance for the integrity verification process. This process is carried out in S_k (i.e.) between the diagnosis center and medical cloud. Let $T_1, T_2, \ldots T_r$ denote the sequence of the medical data transmission (update/store) from the diagnosis center to the cloud. This sequence represents either m_s or m_u such that the decryption of the stored information is carried out, if $m_s + (I \times Q_c) - P_c[I \times (r_1 - n)] = m_s + I[Q_c \times (r_1 - n)] - Q_c[I \times (r_1 - n)]$ or $m_u + [(I \times Q_c) - P_c(I \times r_2)] = m_u + I[Q_c \times (r_2)]$. The above transmission sequence corresponds to the LHS of the transmitted m_s or m_u where in, the RHS is computed with medical cloud proprietary Q_c and P_c. Such integrity verification undergoes a change in update, if $t_p < t_q$. On the other hand, if the change in sequence/update is observed, then, we get the equation as:

$$\left.\begin{aligned}m_s + (I \times Q_c) - P_c \left[I \times \left(r_1 - n - \tfrac{t_p}{t_s}\right)\right] &= m_s + I \left[Q_c \times \left(r_1 - n - \tfrac{t_s}{t_p}\right)\right] - \\ Q_c &\left[I \times \left(r_1 - n - \tfrac{t_s}{t_p}\right)\right] \\ &(or) \\ m_u + \left[(I \times Q_c) - P_c \left(I \times r_2 - \tfrac{(t_q - t_p)}{t_s}\right)\right] &= m_u + I \left[Q_c \times \left(r_2 - \left(\tfrac{t_s}{t_q - t_p}\right)\right)\right]\end{aligned}\right\}$$
(13.5)

In Equation (13.5), the update and store instances for the different sequences of T_1 to T_r is verified for data integrity. This integrity verification is performed in congruence with the access/query time of the end-user. Henceforth, the validation follows different sequence, if faults the above condition, are discarded. This means, the sequence with un-matching integrity is denied for access between cloud and diagnosis center and the user. Thus the freshness and integrity of the medical data is preserved throughout different instances of update and time in the proposed scheme. The verification is adaptive for m_s and m_u independently providing security features in an improved manner.

13.5 Performance Analysis

The performance of the proposed authentication scheme is modeled using OPNET simulator for a varying m size. Information update and store instances are differentiated using message types that are generated. The cloud storage is capable of storing a maximum of 80 Gb data, providing concurrent access to 10 users at a same time. In this simulation, a total of 100 update/store instances are considered for a set of 15 users. The medical information size varies between 256k and 2048 k and the hash generated is 20 for update and store process. With this simulation environment, the proposed scheme is assessed for the metrics process delay, integrity check, and overhead. The proposed scheme is compared for the above metrics with the methods discussed in [8] and [10].

13.5.1 Process Delay

In Figure 13.3, the process delay for the varying hashes is compared between the existing methods and the proposed scheme. The hashes are independently generated for m_s and m_u that differentiates either update or store process without overloading the available time instances. Therefore, the change in

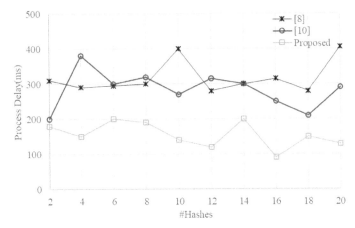

Figure 13.3 Process delay.

hash is generated only if $t_p = t_q$ or $t_p > t_q$ where in only one of m_s or m_u occurs. In particular, the delay for m_u alone is accepted as the number of instances for $m_s < m_u$ and therefore, the process time is less required. The number of varying instances generate r_1 or r_2 notifying the process and hence the delay is less in the proposed scheme.

13.5.2 Integrity Check Bytes

The bytes utilized for integrity check is compared for the methods in [8, 10] and the proposed scheme, as illustrated in Figure 13.4. The number of bytes required for integrity check in the proposed scheme is less due to the preclassification of m_u and m_s. The bytes required for m_u is less than m_s and hence, the update is swift. Similarly, the next sequence for verification is performed on the acknowledgment of r_1 or r_2 as determined by the cloud. The integrity check is initiated from $r_1 + r_2 = 1$ to $r_1 > r_2$ or $r_1 < r_2$ conditions such that required bytes do not consume high for both r_1 and r_2. Preclassification of m_s and m_u and verification of either as in Equation (13.5) reduce the bytes required for integrity check in the proposed scheme.

13.5.3 Overhead

Figure 13.5 presents the overhead analysis of the proposed scheme and the existing methods. The number of verification instances and joint $(r_1 + r_1)$ processing is comparatively less in the proposed scheme. This is due to the

13.5 Performance Analysis 211

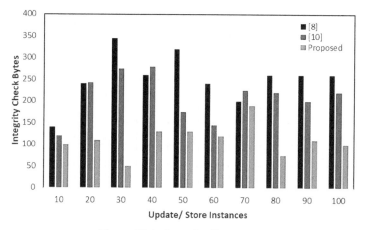

Figure 13.4 Integrity Check Bytes.

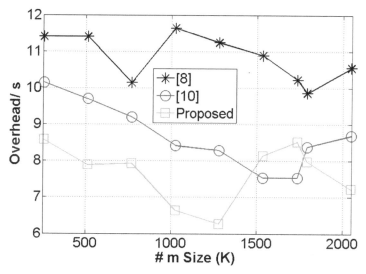

Figure 13.5 Overhead.

differentiation of $t_a = t_r$ and $t_s = \frac{t_a}{t_r}$ at the initial state. Similarly, in generating hashes for different time interval, for the varying m size, either m_s or m_u (for a single record) is performed. The joint validation of process and S_k or S_{ku} determination is prevented in the proposed scheme for varying $T_1, T_2, \ldots T_r$ the differentiation of r_1 and r_2 from $(r_1 + r_2) = 1$ to the incrementing conditions of the authentication scheme helps to reduce the

Table 13.2 Comparative Analysis Results

Metrics	[8]	[10]	Proposed
Process Delay (ms)	409.48	279.85	133.63
Integrity Check Bytes	258	217	96
Overhead/s	10.55	8.69	7.22

overhead with respect to time. In Table 13.2, the comparative analysis results are tabulated.

13.6 Conclusion

This paper introduces an adaptive authentication scheme along with controlled access for medical data stored in cloud platform. This authentication follows conventional elliptical curve cryptography for generating hashes with the precise establishment of session keys. The session keys are adaptive for both diagnosis center to cloud and cloud to user access, retaining the data integrity and availability. The pre-classification of medical data process at the earlier state helps to reduce the generation of unnecessary verification data bytes, requiring less process delay. The performance of the proposed scheme is verified through simulations to verify its consistency through a comparative study.

References

[1] W. Guo, J. Shao, R. Lu, Y. Liu, and A. A. Ghorbani, "A Privacy-Preserving Online Medical Prediagnosis Scheme for Cloud Environment," *IEEE Access*, vol. 6, pp. 48946–48957, 2018.

[2] D. Zheng, A. Wu, Y. Zhang, and Q. Zhao, "Efficient and Privacy-Preserving Medical Data Sharing in Internet of Things With Limited Computing Power," *IEEE Access*, vol. 6, pp. 28019–28027, 2018.

[3] X. Wang, L. Bai, Q. Yang, L. Wang, and F. Jiang, "A dual privacy-preservation scheme for cloud-based eHealth systems," *Journal of Information Security and Applications*, vol. 47, pp. 132–138, 2019.

[4] C.-T. Li, D.-H.Shih, and C.-C. Wang, "Cloud-assisted mutual authentication and privacy preservation protocol for telecare medical information systems," *Computer Methods and Programs in Biomedicine*, vol. 157, pp. 191–203, 2018.

[5] M. J. Guezguez, S. Rekhis, and N. Boudriga, "A Sensor Cloud for the Provision of Secure and QoS-Aware Healthcare Services," *Arabian*

Journal for Science and Engineering, vol. 43, no. 12, pp. 7059–7082, 2017.

[6] K. Chung and R. C. Park, "P2P-based open health cloud for medicine management," *Peer-to-Peer Networking and Applications*, 2019.

[7] S. Barman, H. P. H. Shum, S. Chattopadhyay, and D. Samanta, "A Secure Authentication Protocol for Multi-Server-Based E-Healthcare Using a Fuzzy Commitment Scheme," *IEEE Access*, vol. 7, pp. 12557–12574, 2019.

[8] F. Tang, S. Ma, Y. Xiang, and C. Lin, "An Efficient Authentication Scheme for Blockchain-Based Electronic Health Records," *IEEE Access*, vol. 7, pp. 41678–41689, 2019.

[9] J. Liu, H. Tang, R. Sun, X. Du, and M. Guizani, "Lightweight and Privacy-Preserving Medical Services Access for Healthcare Cloud," *IEEE Access*, vol. 7, pp. 106951–106961, 2019.

[10] R. Ding, H. Zhong, J. Ma, X. Liu, and J. Ning, "Lightweight Privacy-Preserving Identity-Based Verifiable IoT-Based Health Storage System," *IEEE Internet of Things Journal*, vol. 6, no. 5, pp. 8393–8405, 2019.

[11] X. Zhou, J. Liu, Q. Wu, and Z. Zhang, "Privacy Preservation for Outsourced Medical Data With Flexible Access Control," *IEEE Access*, vol. 6, pp. 14827–14841, 2018.

[12] C.-T. Li, D.-H.Shih, and C.-C. Wang, "Cloud-assisted mutual authentication and privacy preservation protocol for telecare medical information systems," *Computer Methods and Programs in Biomedicine*, vol. 157, pp. 191–203, 2018.

[13] H. Zhao, L. Wang, Y. Wang, M. Shu, and J. Liu, "Feasibility study on security deduplication of medical cloud privacy data," *EURASIP Journal on Wireless Communications and Networking*, vol. 2018, no. 1, 2018.

[14] M. Masud and M. S. Hossain, "Secure data-exchange protocol in a cloud-based collaborative health care environment," *Multimedia Tools and Applications*, vol. 77, no. 9, pp. 11121–11135, 2017.

[15] H. Kaur, M. A. Alam, R. Jameel, A. K. Mourya, and V. Chang, "A Proposed Solution and Future Direction for Blockchain-Based Heterogeneous Medicare Data in Cloud Environment," *Journal of Medical Systems*, vol. 42, no. 8, Oct. 2018.

14

E-Tree MSI Query Learning Analytics on Secured Big Data Streams

B. Balamurugan[1] and S. Jegadeeswari[2]

[1]Assistant Professor, Bharathidasan Govt. College for Women, Puducherry, India
[2]Rajiv Gandhi Arts & Science College, Puducherry, India
E-mail: drbalamurugan@dhtepdy.edu.in; jegadeeswari@dhtepdy.edu.in

Abstract

Data stream process on the cloud infrastructure run continuously with the varying load factors. Cloud infrastructure presents the system to meet the fluctuations on computational load. Cloud infrastructures meet the end to end latency objective and effectively predict the data streams of ensemble models. A different type of data stream supports the deep learning processing for all the workflows. Many data stream processing depends on work load balancing and operator scheduling on the secured cloud environment. Cloud computing uses the virtualized processing and storage resources in conjunction with modern technologies where it delivers the conceptual, scalable platforms and applications as on data services. Ensemble Tree Metric Space (E-tree MSI) analytics builds the system with deep learning process on the cloud infrastructure for the faster prediction of the results with effective balancing of load factor. Metric Space Indexing in E-tree MSI technique executes the classification operations on the cloud data streams. The stream applications analyze the temporal relation between secured data stream. The research work is carried out to perform the secured and fast prediction of the data for balancing the load. This technique improves the performance based on the factors such as CPU load rate, and system flexibility, prediction time, in-order searching result probability rate.

Keywords: Cloud Big Data, Data Stream, Load Balancing, Deep Learning, Metric Space Indexing, E-Tree.

14.1 Introduction

Cloud computing utilize the processing and storage resources in arrangement with modern technologies where it distributes on data services. The Cloud infrastructure provides a huge amount of data processing and presents the system to meet the fluctuations on the computational load [12, 24–26]. Cloud infrastructures effectively predict the data streams with load factors of ensemble models in order to reduce the computation. Data stream processes on the cloud infrastructure run continuously with the varying load factors [19–21]. This chapter proposes a combined framework for Secured Data Stream Mapping to Prediction fast load balancing and Fast Similarity Query (FSQ) learning using E-R$^+$TREE. The combination of the three fundamental techniques for constructing our E-tree MSI technique: Fast Predictive Look-ahead Scheduling approach (FPLS) where the scheduling of spatio-temporal data stream files takes place; Parallel Ensemble Tree Classification (PETC) which performs the process of classification operations on cloud data stream; and bilinear quadrilateral Mapping (BQM) process which adds efficiency retrieval of data in cloud infrastructure and implements an efficient construction of the effective load balancing query processing approach. FSQ learning indexing is developed by combination of E-R$^+$tree FSQ method, this method is used to measure the time complexity level. FSQ learning uses an efficient refining technique, where it searches for more similar queries of the user.

The data stream operation is arranged to start at a primary step and develop to forward on balancing the load on the cloud data stream [17, 18] process. Indexing performs the operations on the cloud data stream to accomplish the classification on respective visit of the cloud users. The classification of data stream in cloud decreases the overload factor and the execution time. The indexing algorithm is able to enthusiastically follow deviations in the constant data stream variations. All the relevant variations which have done on the data are respectively updated on the indexing structure. The mapping uses the basic purpose to linearly predict the result from the cloud data storage and management data streams.

This chapter has absorbed about some of the methods that were applied to protect data. This framework was developed to store the data in cloud with secured data format using cryptography technique which is based on block cipher [11].

14.2 Literature Review

Cloud computing uses the virtualized processing and storage resources in conjunction with modern technologies where it delivers the conceptual, scalable platforms and applications as on data services [2, 26, 31, 32]. The secured billing of these data services is straight tied to usage statistics. Distributed cloud data stream processing engines often utilize the inflexible operator allocation strategies [13, 14]. The data stream process are analyzed the temporal relation between secured data in a cloud data streaming, and the methodology is carried out to perform the secured and fast prediction of the data stream through indexing structure for Load balancing [16].

Existing E-tree Indexing Structure is defined in [1] that systematizes all the base classifiers for predicting the result in minimal time complexity. E-tree linearly scans are all about the historic classifiers during prediction using the divide and conquer strategy. E-tree estimation is not extended to the fast prediction of the results on Load balancing. E-Tree indexing structure in [3] addresses the prediction efficient problem on cloud data streams. E-tree indexing does not cover the work with cloud data classification for balancing the load. CPDP scheme in [4] presents an efficient method to select the clients and storage service providers. CPDP provides special functions for data storage and management, whereas the large data streams affect the bilinear mapping operations. Multi-tier application deployments on Infrastructure-as-a-Service clouds as demonstrated in [5] develop a multiple linear regression model to predict application performance but the load balance statistics are not carried out on the cloud data stream. To attain load balance on the cloud data stream, Ensemble Tree Metric Space Indexing (E-tree MSI) technique is proposed in this paper. E-tree MSI technique is implemented on the cloud infrastructure, the application of data streams are partitioned and distributed over different servers. Ensemble tree is developed for the fast prediction of the better performance. E-tree balances the high linear load structure with lesser computational complexity [9, 10].

Outsourced Similarity Search as described in [2] uses the encrypted index-based technique to perform the multiple communications for the user query. Outsourced Similarity Search enables high-query accuracy by supporting the insertion and deletion operation. Outsourced Similarity Search result quality is not up to the grade on the R-tree indexing structure. R-tree, a hierarchical encrypted index as described in [6] encrypted half space range queries by achieving the preferred security-efficiency trade-offs. R-tree indexing performance on outsourced database does not perform the effective indexing with lesser computational complexity.

Trusted offline third party on cloud data stream in [7] employs ringers coupled with secret sharing techniques to provide verifiable and conditional e-payments. Here Payments for Outsourced Computations performs the range query search and produces the result with higher complexity level. Flexible multi-keyword query scheme (MKQE) as illustrated in [8] greatly reduces the cloud protection overhead during the keyword dictionary expansion. MKQE Scheme fails on providing the extra functionalities such as semantic query and fuzzy keyword query on the cloud data stream.

To provide the security environment in cloud on the data query processing, E-R$^+$tree Fast Similarity Query (E-R$^+$tree FSQ) Search is presented. E-R$^+$tree is a combined tree structure constructed for balancing the load and performs the similarity search. E-tree components is used on constructing the E-R$^+$tree FSQ search method. R$^+$tree mechanism carries out the multidimensional indexing with the quasi data objects. The data objects processed are used to attain the higher similarity search results. FSQ Search indexing builds the cloud data stream processing in an efficient arrangement of handling the different types of user queries and produce the result with efficient computational complexity [22, 23].

14.3 Proposed Framework-Secured Framework for Balancing Load Factor Using Ensemble Tree Classification

The chapter presents the load balancing framework and the query learning indexing for implementing a secured cloud infrastructure. The cloud environment is framed with E-Tree Metric Space Indexing technique query learning. The chapter introduces three fundamental techniques for constructing our E-tree MSI technique: Fast Predictive Look-ahead Scheduling approach (FPLS) where scheduling of spatio-temporal data stream files takes place; Parallel Ensemble Tree Classification (PETC) performs the process of classification operations on cloud data stream; and Bilinear Quadrilateral Mapping (BQM) process which supports efficient mapping in cloud infrastructure and implements an efficient construction of effective load balancing query processing approach [15, 16].

The developed is a significant amount to maintain and manage different client data streams. Large-scale server includes more amounts of client data streams for processing with load balanced factor. E-Tree Metric Space Indexing technique on cloud platform comprises of a number of computers,

14.3 Proposed Framework-Secured Framework

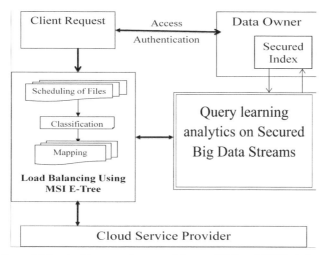

Figure 14.1 Architecture diagram of Secured E-Tree MSI Technique.

resources and store huge number of data streaming. The E-tree MSI technique with load balanced factor accomplishes the client request to resources with minimal execution time and overload factor. The developed cloud platform provides effective indexing technique with large distributed system. The scheduling, indexing, and mapping operation is a significant methodology, which is explained in the architecture diagram of E-tree MSI technique is shown in the Figure 14.1.

Some distributed client system data stream storage in cloud infrastructure is briefly showed in Figure 14.1. The client request to the resources and storage of the data stream is carried out using the E-Tree Metric Space Indexing method. This methodology on the load balancing is carried out through scheduling, classification and mapping process. The tree-based indexing helps to identify the solution for the metric space such as load balance factor, overload factor and execution time. The scheduling process in E-Tree Metric Space Indexing technique is carried out using the FPLS approach. The prediction is carried out in faster manner using the Look-ahead Scheduling process. The cloud servers schedules the files obtained from multiple users, so that the prediction of result in an accurate manner is performed with the balanced load. Next, PETC is used to classify the processed files in tree structure. The tree structure helps to easily perform the mapping process in the final step. The tree structure identifies the relevant aspects and easily predicts the multiple attribute indexing in Ensemble Tree

Metric Space Indexing technique. The mapping is carried out using the BQM process. Bilinear quadrilateral helps to map the client query result in a linear form on the both ends of the tree (i.e., left and right end of tree structure) in minimum time [25, 29, 30].

The new process is added on the cloud data query processing, E-R$^+$tree FSQ Search is presented. E-R$^+$tree is a hybrid tree structure constructed for balanceing the load and performs the similarity search. E-tree components is used on constructing the E-R$^+$tree FSQ search method. R$^+$tree mechanism carries out the multi-dimensional indexing with the quasi data objects. The data objects processed are used to attain the higher similarity search results. FSQ Search indexing build the cloud data stream in arrange of handling the different types of user queries and produce the result with lesser computational complexity.

14.3.1 Fast Predictive Look-ahead Scheduling Approach

The main factor of the FPLS method is to well access the cloud resource for multiple cloud client requests. The query request is handled, initially by scheduling the files. The files are suitably scheduled with the load balanced factor in E-tree MSI technique using the Predictive Look-ahead Scheduling approach. The Request Scheduling using Look-ahead method is handled in an effective way for "i" cloud user at time "t". First, the optimization rule is approved for time period "t" for scheduling the spatio-temporal cloud data stream. The scheduling splits the time interval as "T" denotes the time schedule for "i" users to achieve the adaptive decision making on the cloud data stream. The fast prediction interval strategies set a variation action to maximize the cumulative profit rate with load balanced factor.

The FPLS process in E-Tree MSI Technique, where the current state rules are combined with the following rule to form an effective method with load balanced factor. In cloud different clients' places their request for resources through network communication in cloud environment. Using FPLS, different clients Client$_1$, Client$_2$,...,Client$_n$ place their request for resources like CPU, memory and so on. The cloud server using FPLS procedure in E-Tree MSI Method uses the buffer status table to balance the load in addition to the Look-ahead Data Stream Address Table in the cloud network for efficient scheduling. With this, the Look-ahead-based fast scheduling reduces the regular finish time rate. The objective of FPLS in the E-Tree MSI method supports to increase the utility of the cloud resources with load balanced. As a result, several client requests are scheduled in the cloud in an effective manner followed by which the procedure of classification is achieved in the further section.

14.3.2 Parallel Ensemble Tree Classification (PETC)

In the cloud environment, classification is really required to classify the process of spatio-temporal data stream. In E-Tree MSI method, PETC on the scheduled files recursively organizes for improving the efficiency rate on cloud. The classification of spatio-temporal data stream on tree structure evidently describes the file characteristics. The non-leaf in a tree describes the margins in E-Tree MSI method cloud data stream. The region boundaries support to effortlessly identify multiple characteristics to reduce the overload factor.

PETC with the leaf node on left and right side of the tree in the methodology involves fast prediction in a parallel manner on diverse metric space with modest stable function. To make fast prediction in the PETC technique, the cloud files "F" from the server is processed with modest stable function. File "F" on the cloud environment categorizes the development for "i" users on the left and right side of the tree structure using E-Tree MSI Technique. Parallel Ensemble Learning Rule is approved for the classification of spatio-temporal data stream in E-Tree MSI Technique. The spatio-temporal data stream classification practice in methodology using the parallel ensemble tree procedure. The Parallel Ensemble Learning Rule in E-Tree MSI technique helps to split the data stream on cloud infrastructure in a corresponding manner to reduce the overload factor.

E-Tree classification uses the knowledge method to obtain better predictive performance of spatio-temporal data stream with high flexible structure (i.e., negligible overload factor). The indexing table is used to store the classified process of spatio-temporal data stream for dropping computational complexity. The step-by-step procedures are used to classify the scheduled data stream for easy processing with minimal overload factor. The step-by-step procedures perform the classification procedure based on the data stream file given as input. For each file "F_i" on the cloud environment, classifies the process in an effective method for "i" users on the left and right side of the tree. The non-leaf tree subdivision is defined in the parallel ensemble tree to focus on minimizing the overload factor. Thus, the above parallel ensemble tree algorithm splits the files procedure for relaxed mapping from the cloud environment data storage systems [29, 30].

14.3.3 Bilinear Quadrilateral Mapping

The spatio-temporal data streams are classified in an effective way using PETC, well prepared mapping is achieved to run the requested resources for various users in the cloud. The final stage in the methodology is the design of well-organized mapping on the classified cloud data stream files "F". To accomplish the well-organized mapping process on the cloud environment, the E-tree MSI method practices the BQM. BQM procedure is supported out in E-Tree MSI method for execution of the search process on both side of the ensemble tree (i.e., left and right). The mapping function on the indexed table process information contains the function.

The procedure of mapping with the predictive time interval supports to progress the system accuracy. The bilinear mode of mapping progresses the search process efficiency by minimizing the execution time. BQM analyses the request made by different cloud data stream files for mapping using the index table as given. Time-based mapping is a significant criterion in E-tree MSI method to progress the user query processing on "F" data stream files. The continuous spatio-temporal data are efficient in the cloud storage system, and the effective E-tree MSI method is approved to improve the load balance factor with minimal time factor. The unmapped procedure is detached for improving the user query result fetching based cloud storage data stream. Bilinear quadrilateral transformation with additional scheduling and classification process supports to easily predict the result with minimal execution time. The E-tree MSI technique with bilinear map is a purpose that combines the elements of the left and right tree branch to fetch the result in a well-organized manner. As a result, the tree-based classification is carried out to reduce the execution time factor [29, 30].

14.4 Conclusion

Ensemble Tree Metric Space Indexing (E-tree MSI) technique initially builds the system on the cloud infrastructure for the faster prediction of the results with effective balancing of load factor. Metric Space Indexing in E-tree MSI technique executes the classification operations on the cloud data streams. E-tree Similarity Query learning indexing structure constructed for balancing the load and performs the similarity search. An E-tree component is used on constructing with Similarity Query learning indexing. Fast Similarity Query learning uses the pruning technique and improves the grade of performance of CPU load rate, time complexity level, prediction time, indexing time, and overall performance rate.

Acknowledgments

B. BalaMurugan received his MCA degree from St. Joseph's College, Trichy during 2007, M.Phil in computer Science in the year 2009 and completed Ph.D in Bharathiar University, Coimbatore, India. His research interest includes parallel and distributed computing, and network. Currently working as assistant professor in Bharathidasan Govt. College for women, Puducherry.

S. Jegadeeswari received her M.Sc Computer Science degree from Kanchi Mamunivar Centre For Post Graduate Studies, Puducherry during 2005 and completed her research in Bharathiar University, Coimbatore, India. Her research interest includes parallel and distributed computing, and network. Working as assistant professor in Rajiv Gandhi Arts & Science College, Puducherry.

References

[1] Peng Zhang., Chuan Zhou., Peng Wang., Byron J. Gao., Xingquan Zhu., Li Guo., "E-Tree: An Efficient Indexing Structure for Ensemble Models on Data Streams", IEEE Transaction on Knowledge and Data Engineering, 2011.

[2] Man Lung Yiu., Ira Assent., Christian S. Jensen., and Panos Kalnis., "Outsourced Similarity Search on Metric Data Assets", IEEE Transactions on Knowledge And Data Engineering, Vol. 24, No. 2, February, 2012.

[3] Peng Wang., J. Gao., Xingquan Zhu., Li Guo., "Enabling fast prediction for ensemble models on data streams", International conference on Knowledge discovery and data mining., 2011

[4] Yan Zhu., Hongxin Hu., Gail-Joon Ahn., Mengyang Yu., "Cooperative Provable Data Possession for Integrity Verification in Multi-Cloud Storage", IEEE Transactions on Parallel And Distributed Systems, 2012.

[5] W. Lloyd, S. Pallickara, O. David, J. Lyonb., M. Arabi, K. Rojas, "Performance implications of multi-tier application deployments on Infrastructure-as-a-Service clouds: Towards performance modeling", Future Generation Computer Systems, Elsevier journal, 2013

[6] Bogdan Carbunar and Mahesh Tripunitara "Payments for Outsourced Computations," IEEE Transactions on Parallel and Distributed Systems, Vol. 23, No. 2, February, 2012.

[7] Jianhua Tang, Wee Peng Tay, and Yong Gang Wen, "Dynamic Request Redirection and Elastic Service Scaling in Cloud-Centric Media Networks", IEEE Transactions on Multimedia, 2013
[8] Alessandro Margaraa, Jacopo Urbania, Frank van Harmelena, Henri Bala, "Streaming the Web: Reasoning over Dynamic Data", Journal of Web Semantics., Elsevier journal, 2014.
[9] Gaochao Xu, Junjie Pang, Xiaodong FuJaya, Bharathi Chintalapati, Srinivasa Rao T.Y.S. "A Load Balancing Model Based on Cloud Partitioning for the Public Cloud" IEEE transactions on cloud computing, 2013.
[10] Sidhu, Amandeep Kaur, and Supriya Kinger. "Analysis of load balancing techniques in Cloud computing", International Journal of Computers & Technology 4.2, 737–741, 2013.
[11] Sugumaran, M., B. Bala Murugan, and D. Kamalraj. "An Architecture for Data Security in Cloud Computing", In Proceedings of the 2013 International Conference on Information Technology and Applications, pp. 252–255. IEEE Computer Society, 2013.
[12] V. Gulisano, R. Jimenez-Peris, M. Patino-Martinez, C. Soriente, and P. Valduriez. Streamcloud: An elastic and scalable data streaming system. IEEE TPDS, pp. 2351–2365, 2012.
[13] S. Schneider, H. Andrade, B. Gedik, A. Biem, and K.-L. Wu. Elastic scaling of data parallel operators in stream processing. In IPDPS, pp. 1–12, 2009.
[14] B. Gedik, S. Schneider, M. Hirzel, and K. Wu. Elastic scaling for data stream processing. IEEE TPDS, 2013.
[15] Gaochao Xu, Junjie Pang, Xiaodong FuJaya, Bharathi Chintalapati, Srinivasa Rao, " A Load Balancing Model Based on Cloud Partitioning for the Public Cloud", IEEE transactions on cloud computing year 2013.
[16] Della Valle E, Ceri S, Barbieri DF, Braga D, Campi, A first step towards stream reasoning. In: Future Internet–FIS 2008. Springer, Berlin Heidelberg. pp. 72–81, 2009.
[17] Lanzanasto N, Komazec S, Toma I (2012) Reasoning over real time data streams. 2012.
[18] Sheth A, Henson C, Sahoo SS (2008) Semantic sensor web. Internet Compute IEEE 12(4):78–83, 2012
[19] Dautov R, Stannett M, Paraskakis I, "On the role of stream reasoning in run-time monitoring and analysis in autonomic systems", In Proceedings of the 8th south east European doctoral student conference. Thessaloniki, Greece, 20313.

[20] Valle ED, Ceri S, van Harmelen F, Fensel D, "It's a streaming world! Reasoning upon rapidly changing information". Intell System IEEE transaction Vol. 24(6), pp. 83–89, 2009.

[21] Peng Wang, and Chinya V. Ravishankar, "Secure and Efficient Range Queries on Outsourced Databases Using pR-trees," IEEE International Conference on Data Engineering (ICDE), 2013

[22] Ruixuan Li, Zhiyong Xub, Wanshang Kanga, Kin Choong Yowc, Cheng-Zhong Xuc, "Efficient Multi-Keyword Ranked Query Over Encrypted Data in Cloud Computing," Future Generation Computer Systems., Elsevier journal, 2014.

[23] Subashini and Kavitha, "A survey on security issues in service delivery models of cloud computing." Journal of network and computer applications, pp. 1–11, 2011.

[24] Mell, Peter, and Tim Grance. "The NIST definition of cloud computing", 2011.

[25] B. Hore, S. Mehrotra, M. Canim, and M. Kantarcioglu, "Secure multi-dimensional range queries over outsourced data," The VLDB Journal, vol. 21, pp. 333–358, 2012.

[26] M. Li, S. Yu, N. Cao, and W. Lou, "Authorized private keyword search over encrypted data in cloud computing", In ICDCS. IEEE Computer Society, 2011, pp. 383–392.

[27] B. Hore, S. Mehrotra, and G. Tsudik, "A privacy-preserving index for range queries," in VLDB, pp. 720–731, 2004.

[28] W. K. Wong, D. W.-L. Cheung, B. Kao, and N. Mamoulis, "Secure knn computation on encrypted databases", in SIGMOD Conference, 2009, pp. 139–152.

[29] B.Balamurugan, D. Kamalraj, S. Jegadeeswari and M. Sugumaran, "Secured fast prediction of cloud data stream with balanced load factor using Ensemble Tree Classification", Advances in Computing, Communications and Informatics (ICACCI), 2015 International Conference ICACCI 2015, Page no: 424–430, 2015

[30] B. Balamurugan, D. Kamalraj, S. Jegadeeswari and M. Sugumaran, "A Framework to Data Stream with Balanced Load Factor using Ensemble-Tree in Cloud", International Journal of Applied Engineering Research (IJAER) Volume 10, Number 72 (2015) Special Issues pp. 204–207, 2015.

[31] B. BalaMurugan, D. Kamalraj and M. Sugumaran, "Data Security in Cloud Computing – Issues and Solutions to SaaS", International Journal

of Advance research in Computer Science and software Engineering Volume 4, Issue 4, April-2014.

[32] Sugumaran, BalaMurugan and Kamalraj, "An Architecture for Data Security in Cloud Computing", IEEE World Congress on Computing and Communication Technologies, 2014.

15

Lethal Vulnerability of Robotics in Industrial Sectors

R. G. Babukarthik, Terrance Frederick Fernandez, Sivaramakrishnan, and Aiswariya Milan

Department of Computer Science & Engineering, Department of Electronic Communication & Engineering, Dayananda Sagar University, Bangalore, India
Department of Information Technology Dhanalakshmi Srinivasan College of Engineering & Technology, India
E-mail: r.g.babukarthik@gmail.com; frederick@pec.edu; sivaramkrish.s@gmail.com, aiswariya-cse@dsu.edu.in

Abstract

Over 40 years, human existence has drastically changed due to the advancement of information technology with Artificial Intelligence. Robots have been deployed in the manufacturing sectors and industries, hotels, hospital management, and agriculture sectors. Robots play a major role in social agency, virtual systems, and other autonomous physical systems. The research focuses on the adaptation of this novel technology in the usage of robots in industries and social sectors. The chapter discusses the future research of robots in manufacturing industries and their impact on them and its methodology of innovation used for increasing productivity. Walking robots are to be deployed in hotels, the army, and the education sectors. Furthermore, it states the measure that needs to focus on to avoid cyberattacks on robots.

Keywords: Robots, Artificial Intelligence, Manufacturing Industries, Walking Robots, Cyber Threats.

15.1 Introduction

15.1.1 Robotics' Impact on Manufacturing Industries

The digital transformation has made an impact on the adoption of robotics in manufacturing industries more prominent. Still, many companies are enthusiastic about the adaptation of this technology for increasing productivity, but little concerns exhibit such as workforce effort and cost of transformation. An ample amount is required for the transformation of entrepreneurship especially for newly established firms and existing firms, mainly to increase productivity. The role of labor plays an emerging role in medium- and small-sized enterprises, where much additional labor is required.

15.2 Robotics and Innovation

Proactive innovative approaches like new product marketing, capacity increase, product quality improvement, and capitalizing in innovative processes are required for more innovative firms, and moreover, these firms also need the support of technology resources for their performance. In recent years Artificial Intelligence (AI), Machine Learning (ML), and robotics are connected within the innovation economy, particularly in manufacturing industries [1].

Organization for Economic Co-operation and Development (OECD) estimated that nearly 14% of recent jobs vanished due to automation and moreover, 35% significantly affected because of automation. According to Muro et al. [2] in 2019, stated that automation is an activity that replaces human labor work and those that are done by machines. The primary objectives are to reduce the unit cost and to increase the quality of the product manufactured. The traditional set of qualifications is needed if labor moves from one post to another post, thereby automation plays a trivial role by Chui Manyika et al. in 2016 [3]. It is unclear that which type of workers are getting affected by automation as the automation is linked to strengthening human capital, training of employees, and education stated by Frank et al. in 2019 [4].

- Proposed Hypothesis (H1) states the effects of robots in the performance and productivity of industries. Company size is one of the essential matters to consider during automation.
- Hypothesis (H2) defines an increase in the firms due to the adoption of robotics and those firms that does not adopt robotization.
- Hypothesis (H3) depicts that productivity reaches saturation point even after implementing robotics, further productivity may be increased but it is clear that it is not increased.

- Hypothesis (H4) Large, Small, and Medium Enterprise companies adopt robotization with an increase in human labor cost.
- Hypothesis (H5) describes that the increase in cost reaches saturation once the adoption of robotics is completed.
- Hypothesis (H6) represents the effect of robotization is more advantageous in favor of the environment and in the critical financial environment.

15.2.1 Data Collection

Data is collected from the manufacturing companies in Spain from 1990 to 2015 and processed by the Spanish ministry of public administration and finance. The key advantage of the database is that length of the data is 25 years which includes economic recession and recovery period. Furthermore, analysis of data is performed in detail impact of productivity and evaluation of particular economics on manufacturing industries by Torrent sellers et al. in 2018 [5].

The sample contains 4578 manufacturing firms out of which 3656 (79.86%) are small manufacturing enterprises and 922 (20.13%) are larger enterprises. The sample covered two main areas such as prices, markets, costs, and investment for calculated decision making. Small manufacturing enterprises group is split into three clusters namely cluster 1 (66.0%), cluster 2 (2.7%), and cluster 3 (31.2%). Large firm is split into four clusters, cluster 1 (30.8%), cluster 2 (4.9%), cluster 3 (26.8%), and cluster 4 (37.5%). The smallest cluster contains 45 companies and the largest cluster contains 346 companies.

The finding evidence that H1 hypothesis has adopted robotics and proven to be better in terms of labor productivity performance. Furthermore, robotics empower companies' proficient model of productivity with the provision of H2 hypothesis. The result analysis proved that either large firms or small manufacturing enterprises have benefited from the transition from non-robotics to robotics even though the size of companies differs. So irrespective of size of the companies, the company is benefited for transfer of technology into robotics.

15.2.2 Walking Robots

The field of robotics is explored widely due to the motion, development, and design of mobile robots. Mobile robots are applied in many fields such as military application, industrial use, and space exploration. Thereby, the design and development of mobile robots play an important role in usage. The quadruped mobile robot is considered to be superior among other mobile robots like tracked and wheeled because the quadruped robot is featured with legs and it explores like an animal or a human. Priyarnajan et al. [6], in 2020 surveyed quadruped robots on the basis of environment awareness techniques and their design and development. Among the quadruped robots, the spot is considered to be an intelligent and advanced robot.

Mobile robots gain attention especially in the fields of space, tasks where human effort is a need, and rescue operations. Some of the essential characteristics needed for mobile robots are transverse ability, efficiency, maneuverability, controllability, stability, terrain land, navigation over obstacles, and cost-effectiveness. Quadruped robots are among the best choice legged robots for easily controlling design, and maintenance than compared to the two-legged or six-legged robots. It has gained from biologically stimulated locomotion such as cow, cheetah, and dog achieving speed and environmental movement [8–10].

15.2.3 Various Robot Names and Dimensions

GE walking truck [7], robots with a dimension of length 4 m, width 3 m, and height 3.3 m, it is a three-legged robot. Phony Pony [8] can crawl, creep with a maximum weight of 50 kg, and equipped with two legs. Big Muskie [9] with a height of 48 m and length of 94 m is equipped with four legs. Kumo-I [10] robot is 1.5 m in length which can walk and it is featured with one leg. PV-II [11] robot is 0.9 m in length and 1 m in height, it can walk and it is equipped with three legs, and is of 10 kg in weight. TITAN-XIII [12] with a dimension of 0.21 m length, 0.55 width, and 0.34 height, has the technique of sprawling, and its weight is 5.65 kg, and it has three legs.

COLLIE-II [13] with a length of 0.35 m and 0.38 m height can walk, at a trotting pace. It is equipped with six legs and weighs 0.15 kg. SCOUT-I [14] with a length of 0.2 m and walking techniques of the walk and run, containing one leg and weight of 1.2 kg. SCOUT-II [15] of length 0.55 m, 0.48 m width, and 0.27 height, bound techniques having two legs. WRAP-I [16] with walking techniques of trot and crawl, equipped with three legs and weighs 60 kg. Big Dog-1 and Big Dog-2 [17] with a length of 1.1 m, 0.3 m width, and height of 1 m, the bound technique is used for walking with four legs and weighs 109 kg. Littleton [18] of length 0.3 m containing only three legs and weighs 2.85 kg, Cheetah [85] of height 1.7 m, walking techniques of gallop run with three legs, whereas in the case of Wild Cat [86] with a height of 1.17 m and techniques of trot, bound, and gallop, containing only

Figure 15.1 Spot Robot.

three legs. Spot [87] with a length of 1.1 m, width of 0.5 m, and height of 0.84 m, containing only three legs and weighs 30 kg. The Figure 15.1 gives the visual representation of spot robot with four legs.

Normalized work capacity (NWC) is used for evaluating the quadruped robot, thus NSW relates the Normalized Speed (NS) and hence it is directly proportional. Normalized speed is stated as the ratio of body length and maximum speed. Payload capacity (PLC) deals with the weight of the robot. TITAN-XIII provides PLC of 88% and NWC of 387 % whereas SCALF-1 provides PLC of 65% and NMC of 117%. From the observation, it is clear that TITAN-XIII and SCALF-1 are efficient in performance in terms of a hydraulic and electrical quadruped robot. Controller and actuators play an important role in increasing the robustness of the mobile robot. Moreover, actuators play a vital role in the cost, complexity, and weight of the robots. Quadruped robots play a significant role in incorporating Artificial Intelligence (AI), as these robots synchronize leg movements such as galloping, walking, and trotting. Need to focus on features like memorize, recognize, and learn as the current robot is limited to the visual perception system. Social, physical, and emotional interaction is the need for human–robot interaction. Some of the medium-sized quadruped robots developed by the researchers are BigDog, Spot, and MIT Cheetah.

15.3 Robot Service in Hotels

The tourism industry is becoming more noticeable due to technological advancement in terms of robots and Artificial Intelligence (AI). The customer has many options for interaction with robots and human interactions. Robots service in hotels has a great impact during the current pandemic of COVID-19. The Figure 15.2 represents robot used for hotel service. Robot-staffed hotels have gained an importance among customers due to the global health crisis. Robots continue to be involved in the hospital industry and tourism because advancements in AI are the predictable path for innovation thereby, increase in profitability and efficiency [19]. Consumer considered

Figure 15.2 Robots service in hotels.

various attributes in choosing hotel during traveling decision. The key for marketing is to have deeper knowledge in usage and adoption of this technology (UK Pabi et al. in 2017) [20]. Potential issues and skepticism have been expressed by some of the researchers in the acceptance of robots by Io et al. 2020 [21, 22]. Evaluation of robot in servicing the customer, service robots, quality perception in hotels by Choi et al. 2019 [23] experience of the customer in service robots deployed in hotels by Tung et al. 2018 [24]. Most of the customers prefer service robots in the context of a health crisis (COVID-19). The study reveals whether travelers prefer human interaction or robot service using empirical tests.

Postulate several hypotheses for identifying the robot service such as H1a, H1b, and studies 1A, 1B, 2A, and 2B.

H1a. In the pandemic situation (COVID-19), the estimation of robot staff is likely to be higher (versus lower) as the risk of COVID-19 is high (versus low).

A previous study indicates that human staff is preferred by the customers due to the "hospitableness" as it found to customer loyalty and solidify trust by Chao et al. 2007 [25]. Because of some inconsistency, it also leads to dissatisfaction. The choice of using robot staff by replacing human is higher due to the ongoing COVID-19 pandemic situation. The major reason for this is to maintain social distancing and safety. There are various factors responsible for the transmission of disease; some of them are hygiene practices, social distancing, and safety measures.

15.3.1 Study 1A

The participants were around 134 U.S. citizens out of which 45.5% females were recruited from Amazon MTurk. The participants were informed to assume a hotel which is equipped with robot staff, moreover, from the evaluation, it is clear that robot-staffed hotels have been rated higher with Mean = 4.81 and standard deviation, 1.79. Regression analysis is performed and the result indicates significant support for the latest technology.

15.3.2 Study 1B

In this study, 134 participants were involved, and the participants were asked to assume that they need to travel to the hotel during the situation of COVID-19. Thus more participants choose robot-staffed service with the specification of reducing the risk. Some of the limitations of the proposed robot's services are its diverse functions, as the questionnaire is only limited in terms of functional robots. Some of the functions deployed are front desks, information searching, handling luggage, and cooking. Several influential mediators need to study. Need to focus on dependent variables like trustworthiness, attractiveness, intentions, and service quality. Furthermore, the result needs to identify if responses fluctuate due to psychological attributes. Figure 15.3 shows the bar graphs comparison of number of travelers before

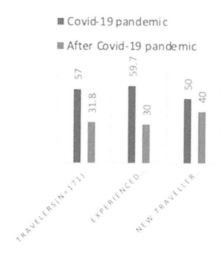

Figure 15.3 Percentage of travelers pandemic.

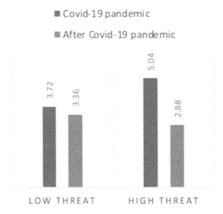

Figure 15.4 Threat due to Covid-19

Covid-19 pandemic and after Covid-19 pandemic. Figure 15.4 depicts comparison graph based on the threat before Covid-19 pandemic and after the Covid-19 pandemic.

15.4 Cyber Security Attacks on Robotic Platforms

The profitability and productivity of the business in the world economy is rapidly transforming due to the usage of robotic technology. A drastic shift in need of optimization and automation in the industry is taking place; it is not only in the manufacturing and warehousing sectors, but also in some of the non-industrial sectors such as farming, defense, schools, hospital, and offices. The major reasons for the new revolution are the availability of open-source platforms, a decrease in electronic prices, hardware, and the merging of technologies. However, potential threats always exist whenever missions and applications are involved.

Robotics are controlled and commanded by applications or by humans. Robots are widely classified as androids, insects, and autonomous. Android robots are intelligent robots which use artificial intelligence and machine learning techniques for learning and reply to the situation on the basis of assigned tasks. Insect performs a function based on a single command, similar to a colony of insects tracked by a single leader. Robotics uses control systems, software applications, networks for messaging, and sensors. However there are lack of standards for various application involving robots. Vulnerability assessments of this technology has to be taken care as they are connected to cloud platform by IoT technique.

Al-shukri et al. [26] proposed the idea of thermal camera image for the robotic system for identification. Tanjim et al. [27] proposed a robotic flight control system for avoiding traffic jams, which reduces traffic jams. The idea of controlling unmanned aerial vehicles is proposed by Uddin et al. [28]. Furthermore, underwater robotic control is proposed by Haus et al. [29] using the vector algorithm. Zhang et al. [30] considered an effective access control model for interactive robots to identify the emotion. Liu et al. [31] traced the vulnerabilities in the Google Play Store application whenever mobile device communication took place and the message is transferred in an unencrypted format. Henceforth leads to a man in the middle attacks and sniffing.

Cyber threats are possible in robots as the robots are made of mechanical parts such as motors, grippers, wheels or pistons and gears, which is very much helpful for lifting, grabbing, and if controlled by malicious people it can cause serious threats. In the month of May 2015, at an United states car factory, a worker's head was crushed by the robot. The reason for this is malware on the robotic platform. In August 2015, in a Maruti factory in India, a robot involved in an assembly grabbed a worker, which led to his death. The reason for this is that the robotic cables were tampered with as they have not followed the security procedure. In April 2016, nine US soldiers died in the US military base as a robot shot the soldiers in the training program and this is mainly due to malfunction in the robot's system. In December 2017, because of misconfiguration in the robotic platform, a toddler ran into a shopping mall and thereby the security flaw was exploited. A similar incident happened in March 2018, when a 3D turtle was attacked and rifled by a robot for the reason that 3D turtle is altered to fool robotic sensors and AI. Cyber threats have become a major challenge in robotics as it has been deployed in various fields such as manufacturing, logistics, agriculture, and healthcare sectors. The approach which was followed in olden days cannot be able to detect this kind of threat. Moreover, confidentiality and integrity have to be maintained as these issues impact greatly in the modern-day sectors. Robots have to detect these risks and mitigate these risks; thereby robots must be cyber safe.

15.5 Conclusion

The advancement in robotics made a tremendous increase in development and productivity in manufacturing industries. Technological advancement in robots using Artificial Intelligence has made the tourism industry booming. The deployment of robots in the hotel industry made an increase in customer satisfaction even in the pandemic situation as there are no indirect human–human interactions involved, but only human–robot interactions. The field of robots is explored due to design, development and motion, hence termed as mobile robots. The article focused on the impact of robots in the manufacturing industries in terms of pros and cons. Robot interacts with humans by helping through trotting, walking, and galloping. The major drawback is to concentrate on the emotional aspect also. The dependent variables that need to focus are attractiveness, trustworthiness, service quality, and intention. The robots should need to be in a position of detecting and mitigate risks associated with it, which in turn leads to cyber safety.

References

[1] Dickson, K.E., Hadjimanolis, A., 1998. Innovation and networking amongst small manufacturing firms in Cyprus. Int. J. Entrepreneurial Behav. Res. 4 (1), 5–17. https://doi.org/10.1108/13552559810203939.

[2] Muro, M, et al., 2019. Automation and Artificial Intelligence: How Machines are Affecting People and Place.

[3] Chui, M., Manyika, J., Miremadi, M., 2016. Where machines could replace human—and where they can't (yet).' McKinsey Q. 30 (2), 1–9. Available at: http://www.oregon4biz.com/assets/e-lib/Workforce/MachReplaceHumans.pdf

[4] Frank, M., Autor, D., J.B., 2019. Toward understanding the impact of artificial intelligence on labor. Proc. Natl. Acad Sci. 116 (14), 6531–6539 doi: pnas.1900949116.

[5] Torrent-Sellens, J., 2018. Las empresas industriales en 2014 y 2015. Encuesta sobre Estrategias Empresariales (ESEE). Fundación SEPI Available at: https://www.researchgate.net/publication/323150903_Las_empresas_industriales_en_2014_y_2015_Encuesta_sobre_Estrategias_Empresariales_ESEE.

[6] Priyaranjan Biswal ⇑, Prases K. Mohanty, Development of quadruped walking robots: A review Ain Shams Engineering Journal 2020

[7] De Santos PG, Garcia E, Estremera J. Quadrupedal locomotion: an introduction to the control of four-legged robots. Springer Science and Business Media; 2007.

[8] Wikipedia. Big Muskie. Available: https://en.wikipedia.org/wiki/Big Muskie.

[9] Hirose S, Kato K. April). Study on quadruped walking robot in Tokyo institute of technology. In: Proceedings of the 2000 IEEE international conference on robotics and automation. p. 414–9.

[10] Kitano S, Hirose S, Horigome A, Endo G. TITAN-XIII: sprawling-type quadruped robot with ability of fast and energy-efficient walking. ROBOMECH J 2016;3(1):8.

[11] Miura H, Shimoyama I, Mitsuishi M, Kimura H. Dynamical walk of quadruped robot (Collie-1). In: Int. Symp.Robotics Research. Cambridge, MA: MIT Press; 1985. p. 317–24.

[12] Buehler M, Battaglia R, Cocosco A, Hawker G, Sarkis J, Yamazaki K. SCOUT: A simple quadruped that walks, climbs, and runs. In: Proceedings 1998 IEEE International Conference on Robotics and Automation, 1998, vol. 2. IEEE; 1998. p. 1707–12.

[13] Battaglia RF. Design of the SCOUT II quadruped with preliminary stairclimbing; 2000. Ingvast J, Ridderströ C, Wikander J. The four-legged robot system WARP1 and its capabilities. In: Second Swedish Workshop on Autonomous Systems; 2002.

[14] Chung JW, Park IW, Oh JH. On the design and development of a quadruped robot platform. Adv Rob 2010; 24(1–2):277–98.

[15] BostonDynamics, CHEETAH – Fastest Legged Robot, Available: http://www.bostondynamics.com/robot_cheetah.html.

[16] J. EMSPAK, WildCat Robot Gallops, Bounds. Available: http://news.discovery.com/tech/robotics/wildcat-robot-gallops-bounds-131004.htm.

[17] Boston Dynamics, Spot. https://www.bostondynamics.com/spot.

[18] Moreda GP, Muñoz-García MA, Barreiro PJEC. High voltage electrification of tractor and agricultural machinery–a review. Energy Convers Manage 2016;115:117–31.

[19] Ivanov, S., Webster, C., 2019a. Conceptual framework of the use of robots, artificial intelligence and service automation in travel, tourism, and hospitality companies. In.

[20] Ukpabi, D.C., Karjaluoto, H., 2017. Consumers' acceptance of information and communications technology in tourism: A review. Telemat. Inform. 34 (5), 618–644.

[21] Lewnard, J.A., Lo, N.C., 2020. Scientific and ethical basis for social-distancing interventions against COVID-19. Lancet Infect. Dis. 20 (6), 631–633.

[22] Io, H.N., Lee, C.B., 2020. Social media comments about hotel robots. J. China Tour. Res.

[23] Choi, Y., Choi, M., Oh, M., Kim, S., 2019. Service robots in hotels: understanding the service quality perceptions of human-robot interaction. J. Hosp. Mark. Manage. 29 (6), 613–635

[24] Tung, V.W.S., Au, N., 2018. Exploring customer experiences with robotics in hospitality. Int. J. Contemp. Hosp. Manage. 30 (7), 2680–2697.
[25] Chao, P., Fu, H.P., Lu, I.Y., 2007. Strengthening the quality–loyalty linkage: the role of customer orientation and interpersonal relationship. Serv. Ind. J. 27 (4), 471–494.
[26] D; Lavanya, V; Sumesh, P; Krishnan, P. 'Intelligent border security intrusion detection using IoT and embedded systems'. 4th IEEE MEC International Conference on Big Data and Smart City (ICBDSC), Muscat, Oman, Feb. 2019.
[27] Tanjim, M; Oishi, A; Nandy, A; Jannah, R; Ahmed, S. 'A flight control system for a vehicle'. IEEE International Conference on Robotics, Electrical and Signal
[28] Uddin, S; Hossain, R; Rabbi, S; Hasan, A; Zishan, R. 'Unmanned aerial vehicle for cleaning the high rise buildings'. IEEE International Conference on Robotics, Electrical and Signal Processing Techniques (ICREST), Dhaka, Bangladesh, 2019.
[29] Haus, T; Orsag, M; Nunez, P; Bogdan, S; Lofaro, D. 'Centroid vectoring for attitude control of floating base robots: from maritime to aerial applications'. IEEE Access, Vol. 7, pp. 16021–16031, 2019.
[30] Zhang, U; Qian, Y; Wu, D; Hossain, S; Ghoneim, A; Chen, M. 'Emotion-aware multimedia systems security'. IEEE Transactions on Multimedia,
[31] Liu, K; Shen, W; Cheng, Y; Cai, L; Li, Q; Zhou, Q; Niu, Z. 'Security analysis of mobile device-to-device network applications'. IEEE Internet of Things Journal, Early Access, 2018.

16

Smart IoT Assistant for Government Schemes and Policies Using Natural Language Processing

J. Pradeep, K. Manojkiran, V.P. Gopi, and B. Jayakumar

Department of Electronics & Communication Engineering, Sri Manakula Vinayagar Engineering College, Puducherry, India
E-mail: pradeepj@smvec.ac.in; mjk1270@gmail.com; vpgopi2505@gmail.com; jayakumar7120@gmail.com

Abstract

Our chapter describes the automatic ways of conveying the government schemes, rules, regulations, and basic policies for the well-being of the people. The beneficial schemes and policies for the people are not able to reach every common man of the country as there is no proper medium of contact. Text-To-Speech, a part of Natural Language Processing has been one of the challenging research fields which aim to develop smart electronic gadgets. The chapter presents a mini computer-based smart assistant with an integrated Text-To-Speech synthesizer for sending the government schemes and policies to the common people as a short message in user preferred languages. The conversion of text form to audio form is developed as equivalent to the natural human voice. This model is emphasized with Natural Language Processing and Digital Signal Processing. The schemes are mainly posted on government websites and portals. Web scraping is a quick and efficient extraction method to fetch data in the form of news from government official websites. The smart assistant can be used for multiple services. The result is tested with different government websites in various text formats and verified in three different languages. The government schemes and policies narrate the information as the natural human voice in different languages.

Keywords: Government Services, Natural Language Processing Based Smart Assistant, Google Text-to-Speech Engine, Optical Character Recognition, GSM.

16.1 Introduction

In this digital era, the government provides various schemes and policies via government official websites. As people are busy with their day-to-day schedule, they might miss out the important information. The beneficiary schemes can go to a single person or intermediates. The paper presents an automated system to fetch the data from the websites to the people directly. Individuals appreciate utilizing Smart Phones because of their usability. They appreciate accomplishing all their work even without touching the Phone [1]. The online platforms available in the phone offer the chance for savages to proliferate bits of hearsay, bogus data, and hypothesis, and to exploit other untrustworthy data to control client assessment [2].

To extract data from a website, an efficient technique called Web Scraping is used. Web Scraping can be a slow process when it is carried out manually. The research studies have been developing several automated solutions in this domain. The methods followed to retrieve web data are Cut/Paste, Document Object Model [3] for structured data extraction in the early period. But the device proposed uses the vernacular ways of navigating, penetrating, and altering the parse tree. It saves programmers hours of work. This approach helps for the automatic extraction of required information from the website. The system utilizes the parsing of HTML and XML tags of the respective website design to fetch the required information [4].

The vital aim of this method is to sort out the information into a natural human voice. The extracted data is sent to the people using the short message service, an efficient method to send the information even when there is the unavailability of the internet. In recent times, there is an extensive change in the Virtual Assistant user experience [5]. Introducing Natural Language Processing (NLP) to the model enhances a complete interactive user experience. The opposite way of Text-to-Speech using NLP is the conversion of speech to text. It is termed as Natural Language Understanding [6].

16.2 Literature Survey

Subhash S. and Srivatsa P. [5] proposed an AI-based voice assistant which recognizes human voice tones and converts the observed information to text.

16.2 Literature Survey

This assistant is made with the help of gTTS (Google Text-To-Speech) Engine which will convert the input language to English. This system will be more beneficial in text-to-speech analysis. As it uses AI technology, the cost of the system will be high. The mentioned assistant produces an audio file and it will be played using the Playsound package of Python programming. It is developed as an equal to voice assistants like Amazon Alexa, Microsoft Cortana, Apple Siri, Google Assistant, etc. gTTS text analysis and Playsound package in Python programming are the essential approach of the projected assistant.

Sangpal R, Gawand T, Vaykar S, and Madhavi [4] proposed a virtual voice assistant named Jarvis with the interpretation of AI Markup Language comprising of gTTS Python-based state-of-the-art technology. Voice tones are developed using Python libraries. Adoption of dynamic base Python pyttsx was done in the proposed work. This projected approach uses Tkinter for GUI development. For NLP, pattern matching, higher-order matching, and extensible mark-up language were introduced in the assistant.

Supriya Kurlekar, Onkar A. Deshpande, Akash V. Kamble, Aniket A. Omanna, and Dinesh B. Patil [9] projected to design a reader tool for blind people using Python, OCR, and gTTS. The reader tool is developed on Raspberry pi 2. It uses Optical Character Recognition technology for the identification of the printed characters using image sensing devices and computer programming. The device uses a pi camera to get a high definition of video and photographs. It converts images of typed or printed text into machine-encoded text. Images are converted into audio output (Speech). The conversion of the printed document into text files is done using Raspberry Pi which again uses PyTesseract library and Python programming. The text files are processed and converted into the audio output (Speech) using gTTS and Python programming language and audio output is achieved. The proposed system utilizes Google text-to-Speech and Command Line Interface (CLI) tool to connect Google Translate's request to the main program.

Kennedy Ralston, Yuhao Chen, Haruna Isah, Farhana Zulkernine A [10] delivered a project which examines and compares three prevalent chatbots API. The proposed system is interacted with voice and it is a multilingual chatbot that can effectively respond to users' attitude, tone, and language using IBM Watson Assistant & Tone Analyzer. The chatbot was estimated using a test case model that was targeted at replying to users' needs. The system was powered with AI for a user-friendly experience so that it resembles a human-like conversation. IBM Watson has a translator with auto-detection of the input languages for up to 24 languages to the system. Categories of chatbots were discussed in the proposed project. Comparative study of several assistants like IBM Watson, Amazon Alexa, and Google Assistant was done. Tone analyzer and Personal Insights of the user were captured by the model. Multilingual voice support was provided by the system.

Si Y., Zhou W., and Gai J. [3] proposed a method to get the data from unstructured Chinese text. A rule-based method built on NLP and regular expression is applied. This system makes use of the linguistic rules of the data in the text and other related knowledge to a proper method. Data representation in the Chinese and English languages was developed by authors. Four main implementation methods of NLP are taken into account for developing this model. Data extraction is divided into two parts: corpus input, word segmentation and rule matching. The designed algorithm is implemented through Python programming in the Windows 10 environment. This chapter will introduce the running process of the algorithm. The text inputs are made in two forms, one is by manual input and the other by getting the list of items as a list. Chinese word segmentation library is utilized from Python Programming. The algorithm for extracting the key to data is from Chinese text. The subprocesses defined are word segmentation, data extraction, and result storage.

Veena G., Hemanth R., and Hareesh J. [12] mentioned the work of creating a relation between different medical data. The data usually contains a lot of unstructured or semi-structured data, by implementing methods like labeling and path similarity analysis. Conversion of unstructured into a structured or classified form was done. Other methods that we use in our work are web scrapping, regular expressions, and part-of-speech tagging. All these methods are implemented in python. Relation extraction of data between different entities is observed from the projected work. Data extraction is done with the help of NLP. NLP processes include Web scraping, regular expressions, part-of-speech tagging, labeling, and path similarity analysis. The method used in the projected system enables extraction of data from the websites.

Stephanie Lunn, Jia Zhu, and Monique Ross [13] explained briefly about Web Scraping and NLP. The Framework of NLP is provided, which adds up more detail to our proposed model. The detailed concepts of web scraping and NLP are explained clearly. This literature analyzes the effective technique of web scraping to extract pedagogical practices from the websites. NLP helps in the text mining of the obtained information. The job market websites are tested in this system. This system paves the idea of getting the required approach for our proposed model.

Shrikrushna Khedkar and G. M. Malwatkar [14] mentioned a model of developing a home automation application using Raspberry Pi and GSM. Programming has been developed in the Python environment for Raspberry Pi operation. A web-based home automation application is proposed in the project. It utilized the Analog to Digital Communication and Global System for Mobile Communications (GSM) along with Raspberry Pi microcontroller. GSM is a wireless system that uses TDMA, most widely used digital wireless technology. It is operated in the 900 MHz to 1800 MHz frequency band in the projected model. GSM modem having a bidirectional connection to the Raspberry Pi used for communication between user and system. This model helped in designing the communicating link between the devices.

Duc Chung Tran, Ahamed Khan M. K. A, Sridevi S [15] the projected work is based on training and testing of end-to-end Text-to-Speech Application. This work presents an approach for automatic data preparation that is used in Tacotron, Tacotron-2-based Mozilla TTS engine. The well-labeled dataset namely FPT Open Vietnamese Speech Dataset having over 25,000 text lines and recorded audio files is demonstrated in this work. Different data sets are produced in .txt formats and .csv formats. FPT Open Speech Dataset (FOSD) approach was used in training the text-to-speech dataset applications.

Partha Mukherjee, Bhowmick S, Paul A, Chatterjee P, & Deyasi A [16] mentioned a Graphical User Interface tool for Text-to-Speech using NLP. It is used for manipulation of the instructions in the text to audio output. Text-to-speech synthesizer converts scripts into speech, by processing them with the help of NLP and Digital Signal Processing (DSP) technologies to convert the textual form into synthesized speech form. Here, the development of a Text-to-Speech synthesizer in the form of a simple application that saves the audio as an mp3 file. It is implemented with a TTS system which converts any text into a human-like voice. It utilized the tree of speech units for better clarity of output tone. Front end and back-end interface for developing the tool for Text-to-Speech Recognition using NLP was proposed in this model. TTS Synthesis model helps in the conversion of Text to speech. The GUI projected here is developed with C# upon. Net Framework 3.5. The feature of conversion of one language to another is not available in the proposed work.

16.3 Proposed Smart System

In this section, the design model of Smart Assistant is defined. The smart assistant includes the processes like Data extraction, Data Processing, Sending of SMS, Language Translation, and Text-To-Speech Approach. The general blocks of the proposed system are shown in Figure 16.1.

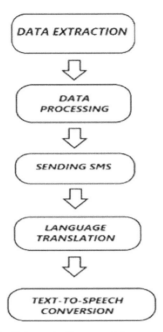

Figure 16.1 Schematic diagram of the proposed system.

16.3.1 Data Extraction

In chapter, the web data has been extracted from the government websites using an effective web scraping technique. The web data can be in various formats such as images, CSV, or PDF files [9]. The textual information from the given data is obtained and made into the required form to transfer it to the people. Extraction of the web data from the website has been done with parsing of tags, and attributes and even multi attributes can be passed to fetch the web data [7].

The data can be in different formats. It can be an image or a PDF file. The admin of the governmental organization can upload the file and send it for processing in a fraction of seconds.

16.3.2 Data Processing

The extracted schemes and policies from the given government are formatted as the needs of the administration. The processing of data is mainly taken into account by the developers of the website. A user-friendly application is developed to extract and send information to people. The message obtained from various forms like jpeg file or PDF file is wrapped as SMS.

16.3.3 Sending SMS

The scheme and policies after data processing the data are obtained as text form. This is delivered to the people as SMS using Application Package Interface (API).

The API acts as a port between the application and mobile number that is interlinked with GSM in the Smart Assistant. It helps to send the text data to the GSM that is deployed in the device. The API utilizes the HTTP request POST method to send the schemes and policies. The website is concatenated along with the schemes and policies extracted from the website, which helps the people for reference.

16.3.4 Language Translation

The Language Translation stage is the prefinal stage of the proposed smart assistant. The language translator helps for viewing the text data in the user's preferred languages. Google API is connected to the Smart Assistant system.

For training a language, the usual method is teaching letter formats and grammatical concepts to the assistant. The language translator discovers the rules of the language from millions and millions of translated text formats automatically. These texts are obtained from books, organizations from all around the world. The translator utilizes an algorithm named Statistical Machine Translation (SMT). SMT is built on language pattern matching. It breaks down each sentence into distinct words or phrases to obtain a match from their datastore. The language translator analyzes a text pattern with the already translated text and gets the text translated into the preferred language. The device supports the bulk translation of the text.

16.3.5 Text-To-Speech

The final stage of this model is Text-To-Speech Synthesis-the conversion of text form to audio as the natural human voice. In this smart assistant, a gTTS engine is installed. The output of the language translator is feed to the Text-To-Speech engine. The engine synthesizes natural language as a human voice in the given language. The process of Text-To-Speech translation is managed by gTTS. The gTTS engine analyzes the linguistics of the input text and relates it with the information on the speech database. The engine saves the audio data in the form of a .mp3 file [2]. As a result, required audio output as human natural language is obtained.

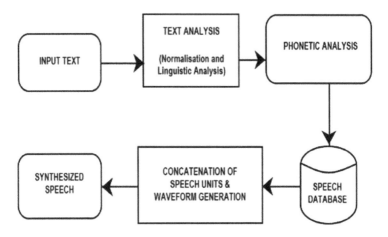

Figure 16.2 Architecture Smart IoT Assistant.

16.3.5.1 Input text

The complete architecture of Smart IoT Assistant is shown in Figure 16.2. The text data is applied as input to the Text-To-Speech engine. The input text is the processed data from the government administration. The text data is obtained from the official website of the governmental organization.

16.3.5.2 Text analysis

The symbolic raw text data is converted to written-out words. The process is called Text Normalization. The text analysis helps in understanding the input text format.

16.3.5.3 Phonetic analysis

The analyzed text is written out as phonetic transcriptions for each word. The text is divided into phrases, clauses, and sentences. Assigning phonetic transcription is called text-to-phoneme or grapheme-to-phoneme conversion [10].

16.3.5.4 Speech database

The processed text containing sentences, syllables, words, phrases, and clauses is specifically revised using a speech recognizer. The unit text is analyzed and framed as speech. The process of segmentation and phonetic acoustics modeling [1] like frequency (pitch), duration is designed and allocated in the database. The required text and the speech audio are forwarded for concatenation.

16.3.5.5 Concatenation & Waveform generation

The process of the grouping of unit speech and producing frequency waveform is the main role of this block. The audio is available in the form of frequency waves and this is forwarded to the synthesizer section.

16.3.5.6 Synthesized speech

There are different forms of speech synthesis. The Smart Assistant uses Deep Learning-Based Synthesis. The speech is delivered as the audio in .mp3 format. It is played by an external speaker unit

16.4 Methodology

The Government Schemes and Policies are available in official web portals and the necessary announcements, rules and regulations to be followed are posted to reach the common people. The Smart Assistant is built with the hardware components like

a Raspberry Pi 3B+
b GSM SIM900A module.
c Matrix Keyboard

The processor used is the Raspberry Pi 3B+. This acts as the main element of the Smart Assistant. The GSM SIM900A is connected to the processor. The data received from the sender is forwarded and processed with the help of the main processor. The interface for the user is the matrix keyboard. Based on the instruction provided by the program, they can select the option at their convenience. The internal circuit design of the Smart Assistant is shown in Figure 16.3.

16.4.1 Input Text Data

The data is obtained from the website using Web Scraping using the Beautiful Soup module with the help of Python Programming. The extracted text data is wrapped and formatted to send it as SMS. The government's official website holds the necessary details that are to be known by the common people. There are three different methods of text data extraction:

- URL Data Extraction
- Image to Text Conversion
- Extract Text from PDF

Figure 16.3 Internal construction of Smart Assistant

16.4.2 URL Data Extraction

Website is given as the input by the government organization; the part of the data from the website is extracted using the class method which is present in HTML backend programming. The data extracted can be in text format without visiting the browser [6]. Text is wrapped up to a limited number of characters. The extracted information along with the website is sent as SMS to the desired users.

16.4.3 Image to Text Conversion

An image consists of text data, can be uploaded by the admin to send the data to the people of the country. PyTesseract library [3] for optical recognition is used to fetch the text from an image. Python programming helps the developer of the governmental organization to get their work at ease. A clear image is required and so it can be processed with the help of the PyTesseract module.

16.4.4 Extract Text from PDF

The smart assistant is developed to extract text from the pdf too. The pdf can be uploaded and the text from it can be sent to the required number of people using API. The text data extraction is done with the help of the

PyPDF2 library of python programming. The obtained text data is ready to reach common people using SMS.

16.4.5 SMS Update

The textual government schemes and policies are sent to the people using the Application Package Interface. The API is for sending the text data to the mobile numbers of the common people of the country. The API has a source and destination number, it transfers the information directly to the people as SMS. Sending of government schemes and policies to the people available in various places at a time is made possible.

16.4.6 GSM

The device comes with Raspberry Pi 3B+ which is interconnected with the SIM900A module. The GSM module is connected with the SIM card of the user choice. This number receives the government schemes and policies related information from the government. The GSM is connected to the general-purpose pins of the Raspberry Pi 3B+ and operated at a frequency of 900 MHz to 1800 MHz [8]. The received data is formatted with the processor and the Text-to-Speech process is initiated.

16.4.7 Language Selection

The people can able to select the language in which they preferred to hear the government schemes and policies. The interface is built with the help of a matrix keyboard which is placed on the top of the device. The people or the user can select the languages with the help of it. Language modeling is done to select the user-preferred languages. Text-to-Speech processing is done after the selection of the user-preferred languages.

16.4.8 Text-To-Speech

The text data is analyzed and converted in speech form with the help of the gTTS package which utilizes NLP as a key source. This produces speech on basis of the Deep-Learning speech synthesis [4]. The fundamental process of analyzing and wave-form generation is done by this approach. gTTS module gets the text data from the SIM900A module and converts it to speech as instructed by the user. gTTS saves the audio file in .mp3 format. The audio output is delivered to the users in the preferred languages.

16.4.9 GUI

The Graphical user interface for the processed text data is displayed in the user-preferred languages. Tkinter Library is utilized in the design of the interface. Audio and Visual output make the Smart Assistant user-friendly and unique

16.5 Experimental Results

The proposed smart assistant has been executed using Python 3.8.8 and Raspberry Pi 3b+. Three different languages with three input formats are tested and executed successfully.

The input text is obtained from an official government website using the Web Scraping technique. The text data is sent to a common people's mobile number which is inserted in the Smart Assistant. The data received by the GSM SIM900A module in the Smart Assistant processes the text data in three different languages. The user can switch the language at their convenience.

The structure of the Smart Assistant is shown in the Figure 16.4. The matrix keyboard is placed on the top for language selection by the user. The microprocessor unit which consists of Raspberry pi 3B+ and GSM SIM900A is laid at the second layer of the Smart Assistant build design.

The user interface of the government administration in Figure 16.5 is for sending the information to the people is done with the help of Python

Figure 16.4 Build design of Smart Assistant.

16.5 Experimental Results

programming and the interface is shown in the Figure 16.6. The admin can select the official governmental website and can able to extract the text data and could able to send the information to the common people using the API.

Figure 16.5 User Interface of the government administration.

Figure 16.6 Output viewed by the user.

INPUT TEXT DATA	pradhan mantri jan dhan yojana hon'ble prime minister announced pradhan mantri jan dhan yojana as the national mission on financial inclusion in his independence day
OUTPUT TEXT DATA – TAMIL	பிரதான் மந்திரி ஜான் தன் யோஜனா மாண்புமிகு பிரதமர் தனது சுதந்திர நாளில் நிதி சேர்ப்பது தொடர்பான தேசிய பணியாக பிரதான் மந்திரி ஜான் தன் யோஜனாவை அறிவித்தார்
OUTPUT TEXT DATA – HINDI	प्रधान मंत्री जन धन योजना माननीय प्रधान मंत्री ने अपने स्वतंत्रता दिवस में वित्तीय समावेशन पर राष्ट्रीय मिशन के रूप में प्रधान मंत्री जन धन योजना की घोषणा की
OUTPUT TEXT DATA – ENGLISH	pradhan mantri jan dhan yojana hon'ble prime minister announced pradhan mantri jan dhan yojana as the national mission on financial inclusion in his independence day

Figure 16.7 A table shows different language output.

The output is obtained in both audio and visual text form in Figure 16.7. The output viewed by the user of the assistant is shown in the figure. The text data is translated and tested in three different languages. The accuracy of the result is about 90% for the test cases given.

16.6 Conclusion

The Smart Assistant for delivering the government schemes and policies from various websites is presented in the paper. The system provides a great human-like voice output to the people in the user preferred languages. The accuracy of the smart assistant system analysis is about 90%. As a result, the proposed smart assistant system will be an automated system for conveying the government's official schemes, policies, rules, and regulations to the common people.

The embedded design of the Smart Assistant helps in delivering the web data directly to people from different sources. It enhances the information delivery to the people using NLP, Text-to-Speech services. The assistant adds up an interactive user interface in different languages.

References

[1] Khattar S, Sachdeva A, Kumar R, & Gupta R, "Smart Home with Virtual Assistant Using Raspberry Pi", 9th International Conference on Cloud Computing, Data Science & Engineering, 2019. India.

[2] Sayef Iqbal, Soon Ae Chun, Fazel Keshtkar, "Using Computational Linguistics to Extract Semantic Patterns from Trolling Data", IEEE 14th International Conference on Semantic Computing (ICSC), 2020, USA.

[3] Si Y, Zhou W, & Gai J. "Research and Implementation of Data Extraction Method Based on NLP", IEEE 14th International Conference on Anti-Counterfeiting, Security, and Identification (ASID). 2020. China.

[4] Sangpal R, Gawand T, Vaykar S, & Madhavi N. Jarvis, "An interpretation of AIML with integration of gTTS and Python", 2nd International Conference on Intelligent Computing, Instrumentation and Control Technologies (ICICICT). 2019.

[5] Subhash S, Srivatsa P. "Artificial Intelligence-based Voice Assistant". International Journal of Science and Engineering Applications IEEE Smart Trends in Systems, Security and Sustainability. Vol. 9 2020, United Kingdom.

[6] Yinghui Huang, Hong-Kwang Kuo, Samuel Thomas, Zvi Kons, Kartik Audhkhasi, Brian Kingsbury, Ron Hoory, Michael Picheny, Leveraging "Unpaired Text Data For Training End-To-End Speech-To-Intent Systems", IEEE International Conference on Acoustics, Speech and Signal Processing (ICASSP), 2020, Spain.

[7] Supriya Kurlekar, Onkar A. Deshpande, Akash V. Kamble, Aniket A. Omanna, Dinesh B. Patil. "Reading Device for Blind People using Python, OCR and GTTS", International Journal of Science and Engineering Applications. 2020.

[8] Kennedy Ralston; Yuhao Chen; Haruna Isah; Farhana Zulkernine, "A Voice Interactive Multilingual Student Support System using IBM Watson," 18th IEEE International Conference On Machine Learning And Applications (ICMLA), 2019. USA.

[9] Si Y, Zhou W, & Gai J. "Research and Implementation of Data Extraction Method Based on NLP". IEEE 14th International Conference on Anti-Counterfeiting, Security, and Identification (ASID). 2020, China.

[10] G, V, R, H, & Hareesh J. Relation Extraction in Clinical Text using NLP Based Regular Expressions. 2nd International Conference on Intelligent Computing, Instrumentation and Control Technologies (ICICICT). 2019 India.

[11] Stephanie Lunn, Jia Zhu, Monique Ross, "Utilizing Web Scraping and Natural Language Processing to Better Inform Pedagogical Practice", Sweden, IEEE Frontiers in Education Conference (FIE). 2020.

[12] Shrikrushna Khedkar, Dr. G. M. Malwatkar. "Using Raspberry Pi and GSM Survey on Home Automation". Mumbai. International Conference on Electrical, Electronics, and Optimization Techniques (ICEEOT), 2016.

[13] Duc Chung Tran, Ahamed Khan M. K. A, Sridevi S, "On the Training and Testing Data Preparation for End-to-End Text-to-Speech Application", Malaysia, 11th IEEE Control and System Graduate Research Colloquium (ICSGRC), 2020.

[14] Vamsikrishna P, Sonti Dinesh Kumar, Shaik Riyaz Hussain, Rama Naidu. K.. "Raspberry PI controlled SMS-Update-Notification (Sun) system" IEEE International Conference on Electrical, Computer and Communication Technologies. 2015.

[15] Diouf R, Sarr E. N, Sall O, Birregah B, Bousso M, & Mbaye, "S. N. Web Scraping: State-of-the-Art and Areas of Application". IEEE International Conference on Big Data (Big Data), 2019.

Index

A

Access control 124, 164, 201, 205, 235
Accuracy 1, 50, 83, 84, 141, 143, 252
AI–IoT-based Future Healthcare 180
ARIMA 47, 50, 51, 54
Artificial Intelligence 15, 61, 122, 188, 231
Authentication 104, 201, 204, 212
Automated Screening 26, 36

B

Bjonteggard Delta Bit Rate (BDBR) 78, 83, 84
Brain Tumor Detection 87, 90, 91, 95

C

Chest X-ray Imaging 26
Cloud 68, 115, 131, 165, 205, 220
Cloud Architectures 115, 146
Cloud Big Data 216
Coding Unit (CU) 78
Complexity 15, 83, 217, 222, 231
Computer-Aided Diagnosis 25, 38, 42
Convolutional Layers 1, 2, 4, 7, 9, 34
Convolutional Neural Network 1, 7, 12, 34, 48, 104
Cyber Threats 227, 235

D

Data Processing 115, 129, 243, 244
Data Stream 160, 215, 217, 222
Deep Learning 1, 80, 103, 123, 188, 215, 249
Deep Neural Networks 42, 58, 104
Depth Intra Coding 77, 82, 83, 84, 85
Depth Size 77, 83

E

E-commerce 48, 61–64, 71
Ensemble Classifiers 116, 146
E-Tree 215, 217–220, 222

F

Facebook Prophet 47, 48, 50, 55–57
Feature Collection 78
Forecasting 47, 48, 151

G

Google Text-to-Speech Engine 240, 241
Government Services 240
GSM 149, 153, 243, 245, 249

H

Humidity Sensor 149, 153, 154

I

Internet Connected Things 180
Internet of Things (IoT) 115, 150, 159, 179, 185, 203
Intrusion Detection System (IDS) 116, 122, 126

L

Load Balancing 215, 217, 218, 219

M

Machine Learning 4, 115, 149, 153, 159, 169, 179, 195
Machine Learning Algorithm 20, 33, 87, 90, 98
Manufacturing Industries 227–229, 235, 236
medical data 48, 115, 146, 202, 205

Medical Imaging 26, 87, 89, 168
Mel-Frequency 103, 104, 105
Metric Space Indexing 215, 217, 219, 222

N
Nano-Robotic Systems 87, 88, 90, 99
Natural Language Processing 4, 15, 62, 169, 239, 240
Natural Language Processing Based Smart Assistant 240
NLP Algorithm 15, 19, 20, 22

O
Optical Character Recognition 240, 241
Oral Squamous Cell Carcinoma 1, 2, 12

P
Predictions 6, 62, 74
Pulmonary 25, 27, 36

R
Recommended System 62, 64

Robots 227–229, 231, 232

S
Signal to Noise Ratio 83, 84, 85, 95
Smart Devices 180
Smart Era 160
Smart healthcare Systems 160, 195
Soil Moisture Sensor 149, 150, 153, 154
Speaker Recognition 103, 104, 105, 106
Speech to Text Conversion 16, 19, 20, 21

T
Temperature Sensor 149, 153, 154, 197
Threshold Limits 116
Tuberculosis 25, 27, 37

V
Voice Recognition 15, 17, 104

W
Walking Robots 227, 229

About the Editors

Dr. Renny Fernandez is an Assistant Professor at the Department of Engineering, Norfolk State University-USA. He graduated from the Department of Electrical Engineering, Indian Institute of Technology Madras in 2010. Prior to joining NSU, Dr. Fernandez served as a visiting assistant professor at University of Indianapolis and Florida International University. Formerly, he worked as a postdoctoral associate with various research groups at University of Utah, Utah State University, Southern Methodist University and University of Virginia. Dr. Fernandez is a multidisciplinary researcher with a focus on wireless sensing platforms for healthcare, pollution control and agriculture. Dr. Fernandez has over 22 Journal publications and 35+ conference presentations and a patent. Email: refernandez@nsu.edu, rennyedwin@gmail.com

Dr. Terrance Fernandez is an Associate Professor at the Department of Computer Science and Engineering, Dhanalakshmi Srinivasan College of Engineering & Technology, Tamil Nadu, India. He is a technical consultant for "B.M. solutions LLC". He received his PhD from Pondicherry Engineering College. Currently, his research focuses on Deep learning employed in medical arenas and Network security in hybrid architectures. In the networking arena, he has worked on characterizing the Internet and to architect the World Wide Web. He co-authored 20 journal publications, 30+ conference presentations and 3 patents with an h-index of 5. He holds professional affiliations to publishing firms like Springer, Hindawi, IGI Global and IET. Email: frederick@pec.edu